Worcestershire
College of Agriculture
Hindlip, Worcester WR3 8SS
Tel: (0905) 51310   Fax: (0905) 754760

50p

**RECOGNITION
AND CONTROL
OF PESTS AND
DISEASES
OF FARM CROPS**

RECOGNITION AND CONTROL OF PESTS AND DISEASES OF FARM CROPS

Ernst Gram, Prosper Bovien and Chr. Stapel

Text by
Frank Hope, NDH, NDArbor

BLANDFORD PRESS
Poole   Dorset

First published in the U.K. 1969

Revised and enlarged second edition 1980

English Language Copyright © 1969 and 1980 Blandford Press
Link House, West Street,
Poole, Dorset, BH15 1LL

ISBN 0 7137 0995 2

Originally published in the Danish language as *Sygdomme og Skadedyr i Landbrugs Afgrøder*
World copyright © 1956, 1969 and 1975 Landbrugets Informationskontor, Copenhagen.

All rights reserved. No part of this book may
be reproduced or transmitted in any form or by
any means, electronic or mechanical, including
photocopying, recording or any information storage
and retrieval system, without permission in writing
from the Publisher.

Printed and bound in Great Britain by Butler & Tanner Ltd., Frome and London

Colour plates printed in Denmark

CONTENTS

Key to Colour Plates	vi
Colour Plates	1-120
Introduction	121
Control Summary Chart	122
Pests, Diseases and Control	123
Plate Descriptions and Control Notes	133
Index of English Names	
Index of Latin Names	

Key to Colour Plates

The 120 colour plates illustrate pests and diseases of the following crops. Descriptive notes appear on page 133 and following.

Cereals	1-28 and 113-117
Grasses	29-32 and 118
Clover, lucerne and legumes	33-48
Mangolds and beet	49-70
Swedes, rape and crucifers	71-89
Potatoes	90-108 and 119
Carrots	109, 110 and 120
Flax	111 and 112

A-B   Blasting, due to adverse growth conditions; in oats and rye respectively.
C   Red oats, cause unknown (mostly found in vigorous marginal plants —
      compare Plate 27, mites).
D   Nitrogen Deficiency in rye (from fertiliser experiments).
E   Normal rye for comparison with D.

A   Barley from very acid soil.
B-D   Phosphate Deficiency, barley.
E-F   Phosphate Deficiency, wheat.

Potassium Deficiency
A   Oats from pot experiments, range of symptoms.
B   Oats from field experiment.
C   Oats from field experiment, just before harvest.
D   Rye from field experiments.
E   Young barley.
F   Older barley.

A   Cyanamide Scorch, barley.
B   Magnesium Deficiency, barley on soil with high potassium content (secondary attack of *Rhynchosporium secalis*).
C   Manganese Deficiency, oats.
D-E   Magnesium Deficiency, barley on soil poor in lime.

Manganese Deficiency (Grey Leaf).
A   Rye.
B   Barley.
C   Oats (see also Plate 4 C).
D   Wheat, March.
E   Normal wheat.
F   Wheat, May, slight symptoms.
G   Wheat, May, severe symptoms.
H   Wheat, April.

A Drought Scorch in barley (sandy soil, June).
B-C Copper Deficiency (white tip) in oats.
D Copper Deficiency (white tip) in barley.
E Scorching after spraying with dinitroorthocresol.

Frost Injury
A   Wheat, March.
B-C   Rye, April.
D   Rye, May (night frost).
E   Rye, June (night frost).
F   Barley, June (night frost).
G   Rye, May.
H-I   Rye, April.
J   Oats, May.
K-L   Barley, May.

A-B	Borax injury in barley, May and July respectively.
C	Ribbon leaf (genetical abnormality), barley.
D	Iron Deficiency in barley, June.
E-F	Sodium chlorate injury in oats and barley respectively, July.
G	'Physiological speckle' in wheat, June (cause unknown).
H	Ion injury, barley from pot experiment (may be induced by high amounts of fertiliser, low pH, and other causes).
I	Sea-salt injury in barley, May.

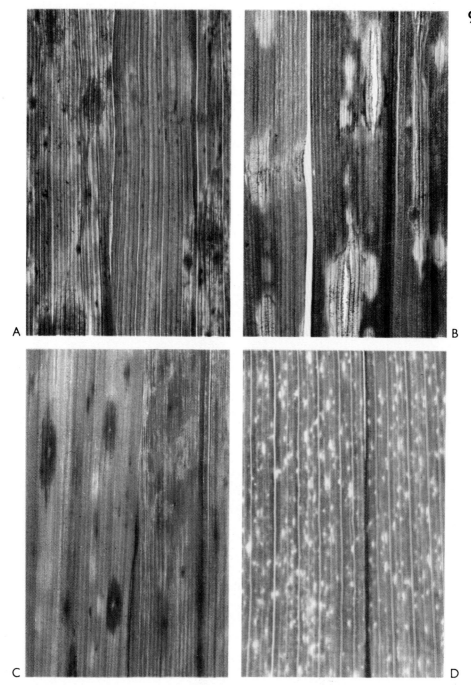

A Halo Blight *(Pseudomonas coronafaciens)* in oats, transmitted light (ca. × 2 mag.). Compare Plate 13 D-F.
B Septoria leaf-spot *(Septoria avenae)* of oats, with numerous pycnidia in the spots (ca. × 2 mag.).
C Dark leaf-spots in oats, cause unknown (ca. × 2 mag.).
D 'Physiological speckle' in wheat, cause unknown (ca. × 2 mag.). Compare Plate 8 G.

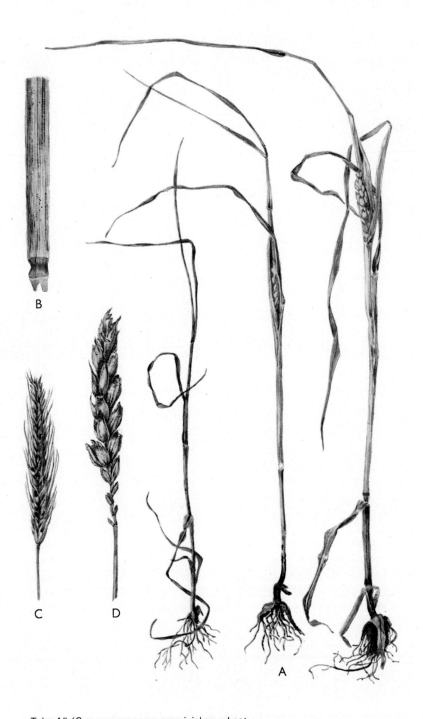

A	Take-All *(Gaeumannomyces graminis)* on wheat.
B	Black Mould *(Cladosporium sp.)* on straw of wheat killed by Take-All.
C	Black Mould *(Cladosporium sp.; Alternaria sp.)* on dead styles, rye.
D	Black Mould *(Cladosporium sp.)* on ear of wheat killed by Take-All.

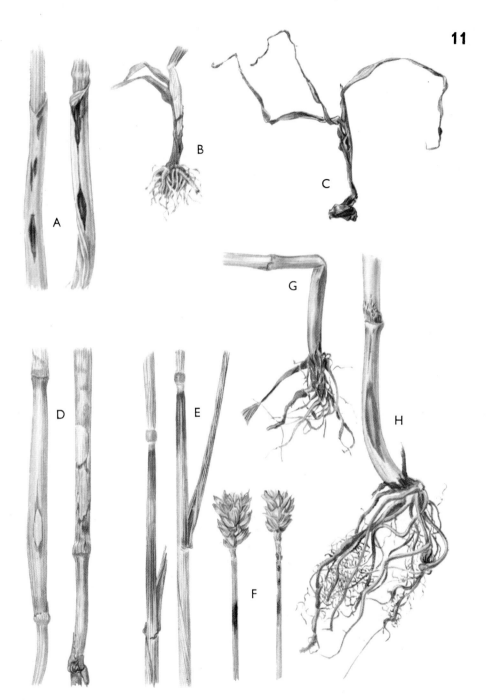

A	Eyespot *(Pseudocercosporella herpotrichoides)*, early infection of wheat coleoptiles (× 3 mag.).
B-C	Eyespot in wheat at the end of winter.
D	Sharp Eyespot *(Rhizoctonia cerealis. Corticium solani).* Material from the Department of Plant Pathology, Rothamsted.
E-F	Melanism in wheat (cause unknown).
G-H	Eyespot *(Pseudocercosporella herpotrichoides)* lodging in rye and wheat, respectively.

A	Brown Foot Rot *(Fusarium spp.)*.
B-C	Seed disinfection injury (overdosage of mercury compounds) in wheat and oats respectively.
D	Ear Blight *(Fusarium spp.)* in rye.
E	Ear Blight *(Fusarium spp.)* following frost injury, wheat.
F	'Oat Cap' *(Fusarium spp.)* on wheat stubble.
G	Snow Mould *(Fusarium nivale),* in wheat.

A-B Leaf-Spot *(Septoria tritici)* on wheat.
C   Leaf-Spot, wheat, January (× 2 mag.).
D-F Halo Blight *(Pseudomonas coronafaciens)* in oats. D recent infection; E shown in transmitted light. Compare Plate 9.
G   Mildew *(Erysiphe graminis)*.
H   Mildew on wheat ear.
I   'Honeydew', conidial stage of Ergot *(Claviceps purpurea)*.
J   Ergot in rye.
K   Ergot with fructification.

A-C	Barley Leaf Stripe *(Pyrenophora gramineum).*
D	Net Blotch *(Pyrenophora teres)* of barley.
E	Leaf-Spot *(Pyrenophora avenae)* of oats.
F	Leaf Blotch *(Rhynchosporium secalis)* of rye.
G	Leaf Blotch *(Rhychosporium secalis)* of barley. Compare Plate 4 B.

A-B	Crown Rust *(Puccinia coronata)* on oats. In B black winter spores.
C	Brown Rust *(Puccinia hordei)* on barley.
D-E	Brown Rust *(Puccinia recondita F. sp. secalis)* on rye.
F	Brown Rust *(Puccinia recondita F. sp. tritici)* on wheat.
G-H	Yellow Rust *(Puccinia striiformis)* on wheat.
I	Black Rust *(Puccinia graminis)* on couch grass *(Agropyrum repens)*.
J-K	Black Rust *(Puccinia graminis)* on rye, summer and winter spores respectively.
L	Black Mould *(Cladosporium sp.)* on dead rye straw.

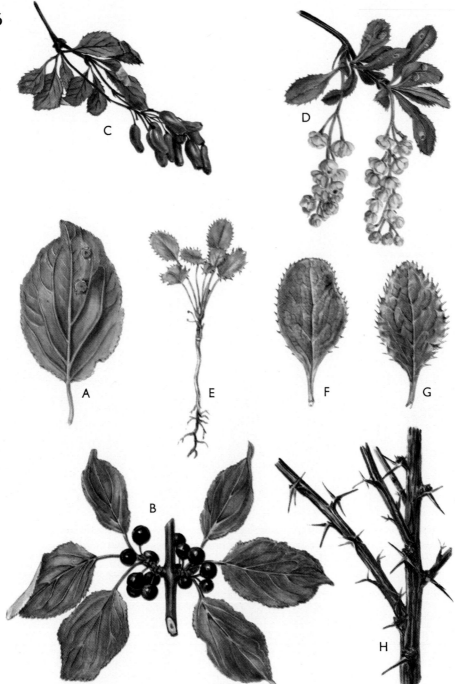

A   Common Buckthorn *(Rhamnus cathartica)*, leaf with cluster cups of Crown Rust. Compare Plate 15 A-B.
B   Common Buckthorn with fruits.
C-D   Common Barberry *(Berberis vulgaris)*, with flowers and fruits respectively. In D cluster cups of Black Rust. Compare Plate 15 I-K.
E   Seedling of Common Barberry.
F-G   Leaves of Common Barberry, showing the 10-30 fine marginal bristled teeth characteristic of the species.
H   Branch of Common Barberry showing triple spines.

A Stripe Smut *(Urocystis occulta)* of rye.
B-E Bunt *(Tilletia caries)*. B smutted head; C bunted grain; D wheat grain infected with bunt spores; E normal kernels.
F Loose Smut *(Ustilago tritici)* in wheat.

A   Covered Smut *(Ustilago hordei)* on barley.
B-C   Loose Smut *(Ustilago avenae)* on oats, early and later stage.
D   Loose Smut *(Ustilago nuda)* on barley, 3 stages.
E   Loose Smut in barley, ear partially infected.

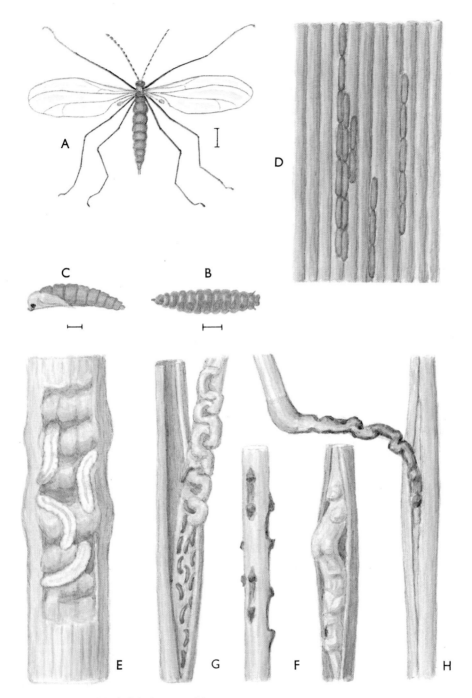

Saddle Gall midge *(Haplodiplosis equestris)*
A   Saddle Gall midge (female).
B   Full grown larva.
C   Pupa.
D   Eggs on a leaf of barley (magnified; length of the egg 0.4-0.5 mm).
E   Young larvae.
F   Saddles on a stem of barley.
G   Older larvae.
H   Destroyed and broken stem of barley.

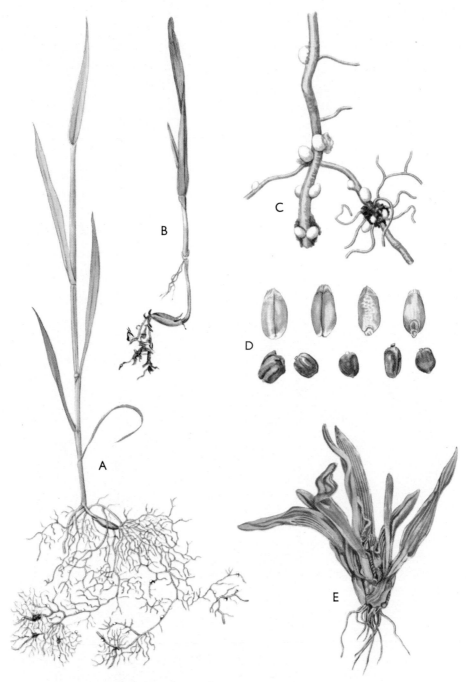

A	The Cereal Root Eelworm *(Heterodera avenae)*. Oat plant with abnormal branching and females on the roots.
B	The Cereal Root Eelworm, attack in early stage.
C	The Cereal Root Eelworm, root with white females attached (ca. × 10 mag.).
D	Normal wheat grains (above), and 'ear cockles' or 'peppercorn' due to attack by *Anguina tritici* (below).
E	The Stem and Bulb Eelworm *(Ditylenchus dipsaci)*, stem disease in young rye plant.

A-B Aphid injury on rye and barley respectively.
C   The Grain Aphid *(Macrosiphum avenae)* on oats ($\times$ 1½ mag.).
D-E Injury on rye and oats caused by Thrips *(Limothrips spp.)*.
F   Barley attacked by the Cabbage Thrips *(Thrips angusticeps)*.

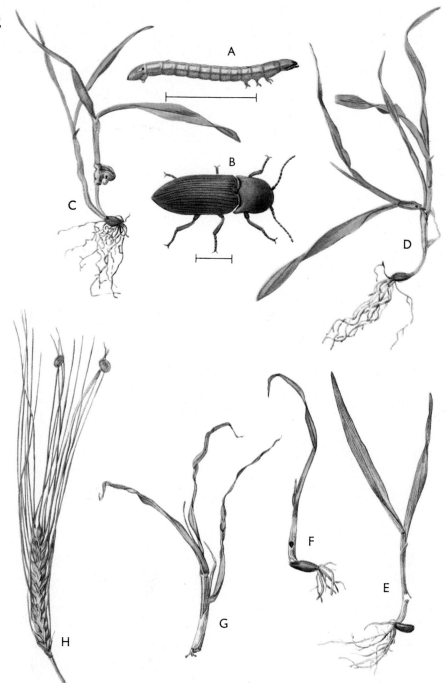

A	Wireworm *(Agriotes sp.)*.
B	Click-beetle *(Agriotes lineatus)*.
C-E	Wireworm injury, C-D barley, E wheat.
F	Barley pierced by larva of the Wheat Flea Beetle *(Crepidodera ferruginea)*.
G	Barley cut by larva of Cockchafer *(Melolontha melolontha)*.
H	Pupae commonly found in ears of barley, belonging to *Phytonomus arator* a curculionid living on weeds.

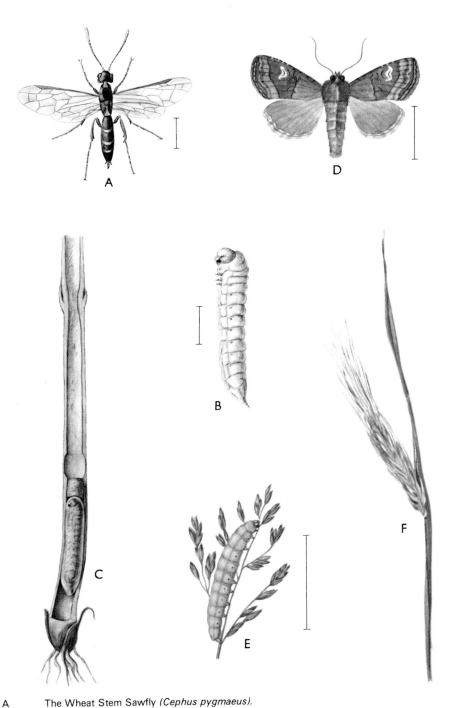

A	The Wheat Stem Sawfly *(Cephus pygmaeus)*.
B	Larva of the Wheat Stem Sawfly.
C	Longitudinal section of wheat stem, showing the cocoon at the bottom of the straw.
D	The Common Rustic Moth *(Mesapamea secalis)*.
E	The caterpillar of the Common Rustic Moth.
F	Whiteheads in rye, the stem cut by the caterpillar of the Common Rustic Moth.

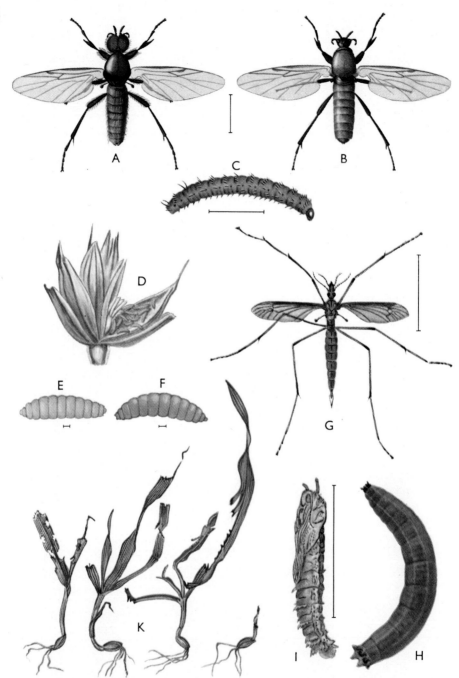

A-B	Bibionid Flies *(Bibio hortulanus)*, male and female respectively.
C	Larva of *Bibio hortulanus*.
D	Larvae of Wheat Midge in spikelet of wheat.
E	Larva of Wheat Midge *(Contarinia tritici)*.
F	Larva of Wheat Midge *(Sitodiplosis mosellana)*.
G	Crane Fly *(Tipula paludosa)*.
H	Larva of Crane Fly ('Leatherjacket').
I	Pupa of Crane Fly.
K	Barley plants damaged by leatherjackets.

The Frit Fly *(Oscinella frit)*.
A-B Blasted spikelets, A from earlier, B from later attack.
C-D Larva and pupa, respectively.
E   Wheat plant with main shoot killed, November.
F   Oat plant with main shoot killed.
G   Late-sown oats with killed main shoots and bulbous bases of tillers.

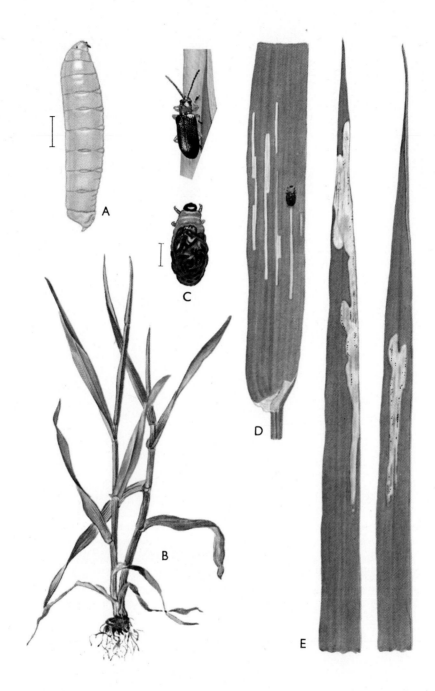

A   Larva of the Wheat Bulb Fly *(Leptohylemyia coarctata)*.
B   Rye plant injured by the Wheat Bulb Fly.
C   Larva of the Cereal Leaf Beetle *(Lema melanopus)*.
D   Barley leaf peeled by larvae of the Cereal Leaf Beetle.
E   Leaves mined by Leaf Miners *(Hydrellia griseola)*.

The Oat Spiral mite *(Steneotarsonemus spirifex)*
A-B   Shoot and inflorescence of oats discoloured red from mite attack.

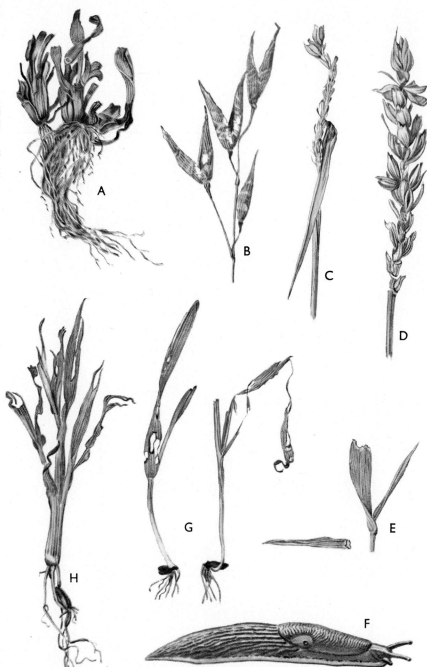

A   Rye plants, bitten by mice *(Microtus agrestis)*, April.
B-D   Injury from sparrows *(Passer domesticus)* in oats, barley, and wheat. In B distinct markings of the beaks.
E   Young barley plant severed by sparrows.
F   Grey Field Slug *(Agriolimax agrestis)*.
G-H   Slug injury.

A-B Frost injury on leaves of cocksfoot.
C   Mildew *(Erysiphe graminis)* on brome grass.
D-E Ergot *(Claviceps purpurea)* on rye-grass.
F-G Leaf Fleck *(Mastigosporium sp.)* on timothy.
H-I Choke *(Epichloe typhina)* on cocksfoot.

A   Yellow Slime *(Corynebacterium rathayi)* on cocksfoot. Compare H below.
B-C   Smut *(Ustilago perennans)* in tall oat grass.
D   Spike-like panicle of timothy injured by the Timothy Fly larvae *(Amaurosoma flavipes)*.
E   The Leaf Beetle larvae *(Lema melanopus)* feeding on the leaves of timothy.
F   Root of rye-grass with cysts of the Cereal Root Eelworm *(Heterodera avenae)*.
G   Ear Smut *(Ustilago bromivora)* on seeds of corn brome grass.
H   Yellow Slime on seeds of cocksfoot. Compare A above.

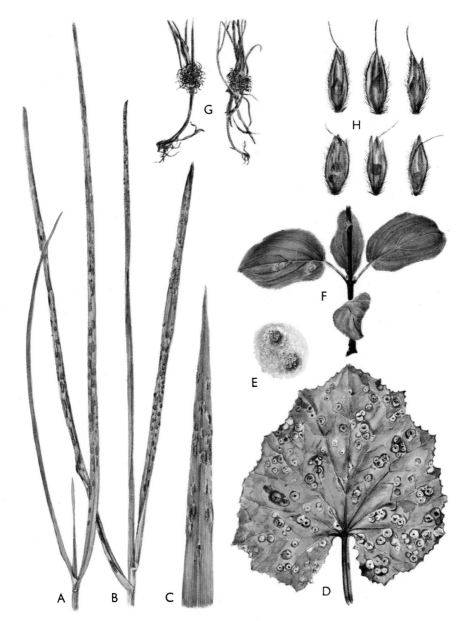

A	Crown Rust *(Puccinia coronata)* on rye-grass.
B	Meadow Grass Rust *(Puccinia poarum)*.
C	Brown Rust *(Puccinia recondita)* on brome grass.
D-E	Aecidial stage of Meadow Grass Rust *(Puccinia poarum)* on the leaf of coltsfoot, E showing aecidia from below.
F	Common Buckthorn with two aecidia of Crown Rust. Compare A above.
G	Red fescue with galls caused by Gall Midge larvae *(Mayetiola sp.)*.
H	Seeds of Meadow Foxtail (in transmitted light). Above: normal seeds; below: destroyed by the Meadow Foxtail Gall Midge larvae *(Dasyneura alopecuri)*.

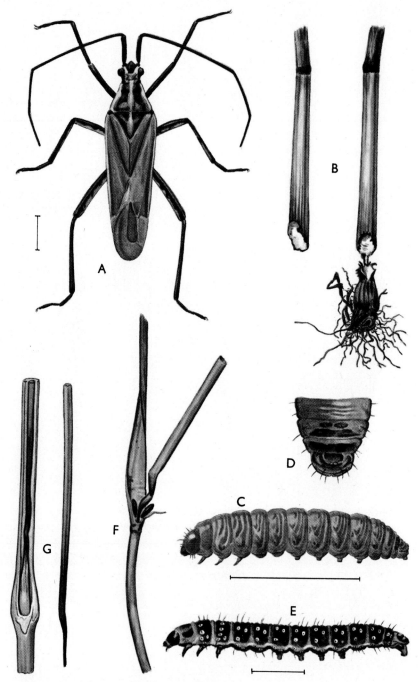

A   The Meadow Plant Bug *(Leptopterna dolobrata)*.
B-D   The Flounced Rustic Moth *(Luperina testacea)*. B Straw of timothy destroyed by the larvae. C the larvae. D the hind part of the larvae (magnified).
E   The Timothy Tortrix Moth *(Tortrix paleana)*.
F   The Hessian Fly *(Mayetiola destructor)* pupae ('flax seeds') between the straw and the sheath.
G   Tarsonemid Mites *(Siteroptes graminum)*. Straws of red fescue grass destroyed by the mites.

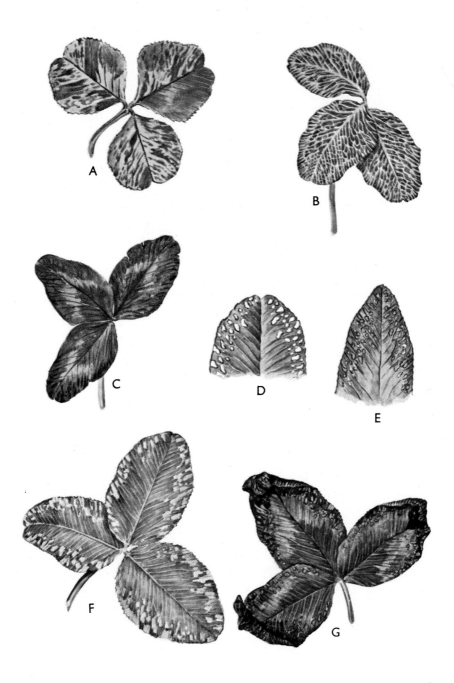

A-B Chlorosis on leaves of clover, possibly due to virus (mosaic). Compare Plate 34A.
C   Necrosis on the surface of a clover leaf caused by the wind.
D-G Potassium Deficiency symptoms. D white clover, F alsike, E and G red clover, with early and advanced symptoms respectively.

A   Mosaic disease on red clover, produced experimentally with virus from field bean.
B-C   Magnesium Deficiency symptoms on white clover and red clover respectively (partly from Wallace).
D   Black Blotch *(Polytrinchium trifolii)* on red clover.
E   Ring Spot *(Stemphylium sarciniforme)* on red clover.
F   Scorch *(Kabatiella caulivora)* on red clover, a stem spot magnified.

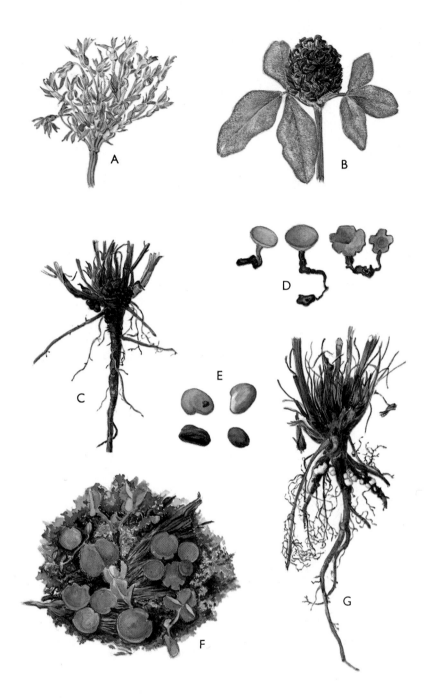

A   Proliferation in white clover, clusters of green leaves appearing on the flower head.
B   Mildew *(Erysiphe polygoni)* on leaves of red clover.
C-G   Clover Rot *(Sclerotinia trifoliorum)*. C and G dead red clover plants with black or white sclerotia on the crowns, E sclerotia in comparison with two red clover seeds, D apothecia emerging from sclerotia, F apothecia seen from above.

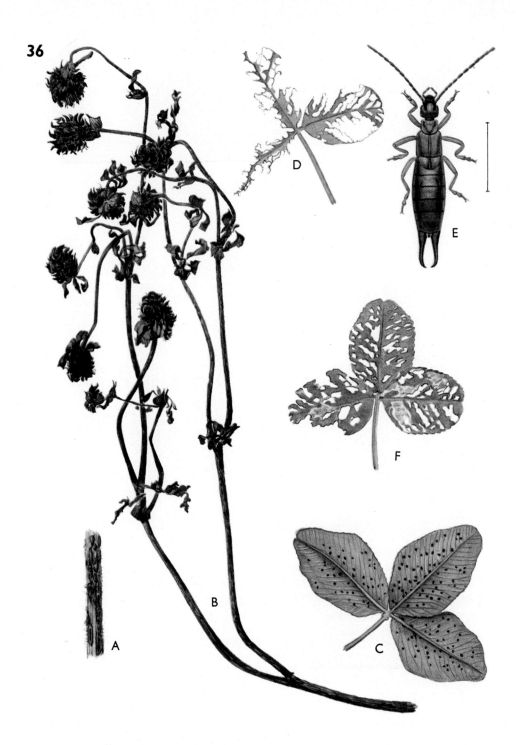

A-C   Rust *(Uromyces trifolii)* on red clover.
D-E   The Common Earwig *(Forficula auricularia)* and typical earwig injury on a leaf of clover.
F   Leaf tissue stripping off caused by the Grey Field Slug *(Agriolimax agrestis)*.

A-B   Swelling and stunting of red clover plants, caused by the Stem and Bulb Eelworm *(Ditylenchus dipsaci)*.
C-D   White clover attacked by the Stem and Bulb Eelworm.

A	Clover Seed Weevil *(Apion apricans)*.
B	Flower head of red clover, split open and showing the damaged florets, caused by Clover Seed Weevil larvae.
C	A white clover floret destroyed by the White Clover Seed Weevil *(Apion flavipes)*.
D	Seeds of white clover gnawed by larvae of the White Clover Seed Weevil.
E	Leaves of white clover attacked by the White Clover Seed Weevil.
F-G	The Clover Leaf Weevil *(Hypera nigrirostris)* and a shoot of red clover destroyed by the larvae.

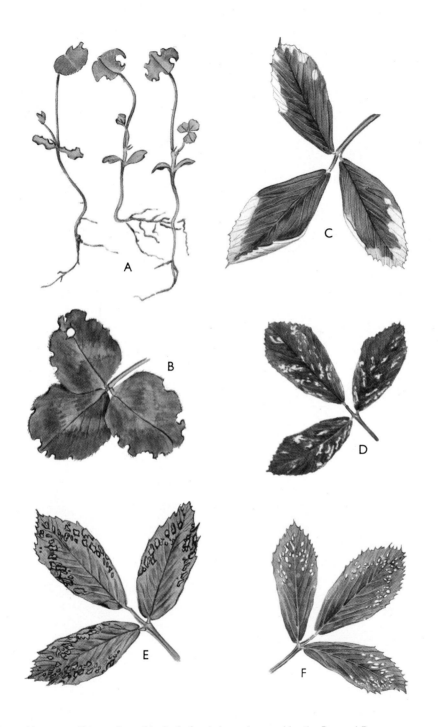

A-B	Young seedlings and an older leaf of red clover damaged by the Pea and Bean Weevil *(Sitona lineatus)*. See Plate 46 A-B.
C	Frost injury on a leaf of lucerne.
D	White spotted leaf of lucerne, probably due to root damage.
E-F	Potassium Deficiency symptoms on leaves of lucerne. Early and advanced stage.

A-B Lucerne with symptoms of Boron Deficiency.
C-E Anthracnose *(Colletotrichum trifolii)* on lucerne, D and E showing stem lesions magnified.
F-H Verticillium Wilt *(Verticillium albo-atrum)* on lucerne, F showing a wilted stem, G and H showing the brown coloured vascular system in the tap root.

A-B Leaves of lucerne with spots caused by the Black Stem Fungus *Ascochyta imperfecta.*
C-D Downy Mildew *(Peronospora trifoliorum)* on lucerne, D showing the undersurface of a leaflet with the violet-grey weft of the fungus.
E   Trefoil (seen from below) destroyed by Clover Rot *(Sclerotinia trifoliorum).* Among the whitish and black irregular and large *Sclerotinia* sclerotia some smaller and spherical sclerotia belonging to the fungus *Typhula trifolii.*
F-G Leaf Spot *(Pseudopeziza medicaginis)* on leaves of lucerne. Early attack on leaflets, and older, withering leaf respectively.
H   Flowers of *Lotus corniculatus* deformed to swollen galls by Gall Midge larvae *(Contarinia loti).*

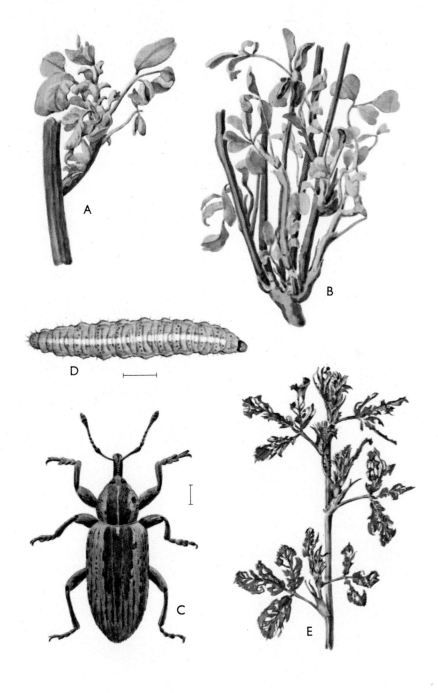

A-B The Stem and Bulb Eelworm *(Ditylenchus dipsaci)* giving rise to swelling and stunting of the shoots of lucerne.
C-E Lucerne Leaf weevil *(Hypera variabilis)*. The weevil, the larva and a plant of lucerne seriously damaged by the feeding larvae.

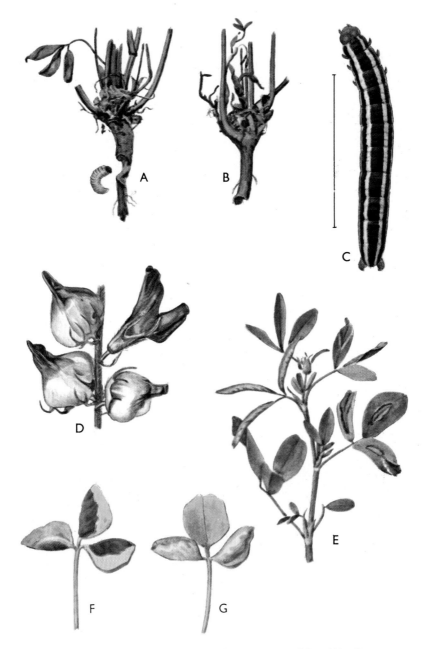

A-B Roots of lucerne plants attacked and gnawed by the larvae of Root Weevils (Otiorrhynchus ligustici).
C The Pea Noctuid Moth larva (Mamestra pisi).
D Lucerne flowers deformed by the Lucerne Flower Gall Midge larvae (Contarinia medicaginis).
E Leaves of lucerne attacked by the Lucerne Leaf Gall Midge larvae (Jaapiella medicaginis).
F-G Leaves of trefoil deformed by Gall Midge larvae (Dasyneura onobrychidis).

A   Foot Rot (unknown origin) on a pea plant.
B   Mildew *(Erysiphe polygoni)* on pea leaves.
C   Grey Mould *(Botrytis cinerea)* at the tip of a pea pod.
D-F   Leaf and Pod Spot *(Ascochyta pisi)* on pods, leaves and stems of pea.

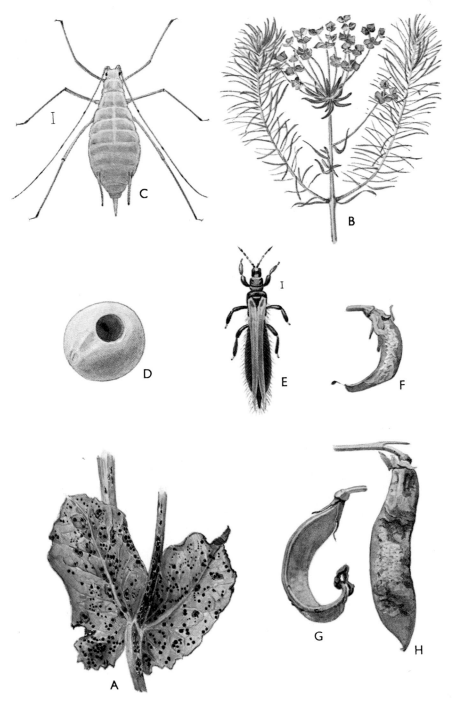

A	Rust *(Uromyces pisi)* on stems and leaves of pea.
B	*Euphorbia cyparissias,* host for the aecidial stage of the rust *Uromyces pisi.*
C	The Pea Aphid *(Acyrthosiphon pisum).*
D	Pea seed showing attack by the Pea Beetle *(Bruchus pisorum).*
E-H	The Pea Thrips *(Kakothrips robustus)* and three pea pods distorted and showing silvery discoloration after attacks of the thrips.

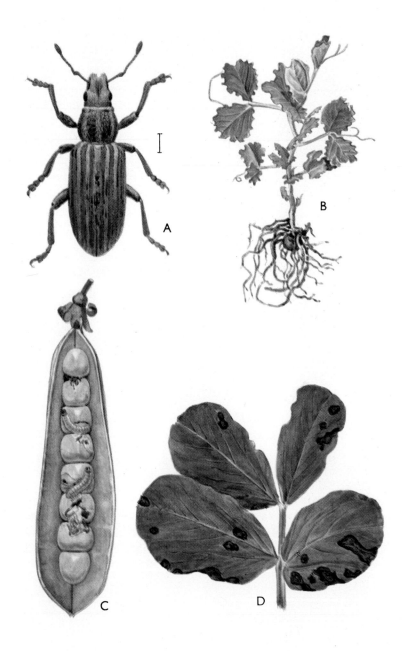

A   The Pea and Bean Weevil *(Sitona lineatus)*.
B   Half-circular notches on the edge of pea leaves, typical feeding habits of the Pea and Bean Weevil.
C   The Pea Moth larvae *(Laspeyresia nigricana)* eating the seeds in a pea pod.
D   Leaf of field bean showing leaf spots caused by the fungus *Ascochyta hortensis*.

A   Grey Mould *(Botrytis cinerea)* on the stem of lupin.
B-E   Attacks by *Ceratophorum setosum* on leaf, stem and pods of lupin.
F   Seeds of white lupin, to the left healthy, to the right two seeds destroyed before harvest by rain and moisture.

A	Root Rot on blue lupin.
B	Mosaic of field bean.
C	The Black Bean Aphid *(Aphis fabae)* on field bean.
D-E	Leaves of lupin deformed by the Black Bean Aphid attack.
F-G	Injury to young lupin plants by the Lupin Seed Fly larvae *(Hylemya florilega)*.

A-B   Potassium Deficiency in mangold, early and advanced symptoms respectively.
C   Leaf stalk of mangold showing yellow colour and brown streaks as a result of potassium deficiency.
D-E   Nitrogen Deficiency in leaves of mangold, D is a prematurely withered outer leaf.
F-G   Magnesium Deficiency in sugar beets, G advanced stage with secondary attack by *Alternaria sp.*
H   Leaf of mangold from plots fertilised with calcium nitrate; corresponding plots with sodium nitrate were normal.

A   Boron Deficiency (Heart Rot) in mangold. First year root with black heart leaves. Compare deficiency symptoms in the Plate 51 A-B.
B-D   Boron Deficiency in seed beet plants, B showing a dead shoot tip, C a damaged seed stem with five new shoots and D a stem with brown discoloration.

A-B   Boron Deficiency, 'Dryrot', two sides of the same sugar beet.
C-D   Sections of the same sugar beet as in A-B.
E   Mangold with pronounced Boron Deficiency.

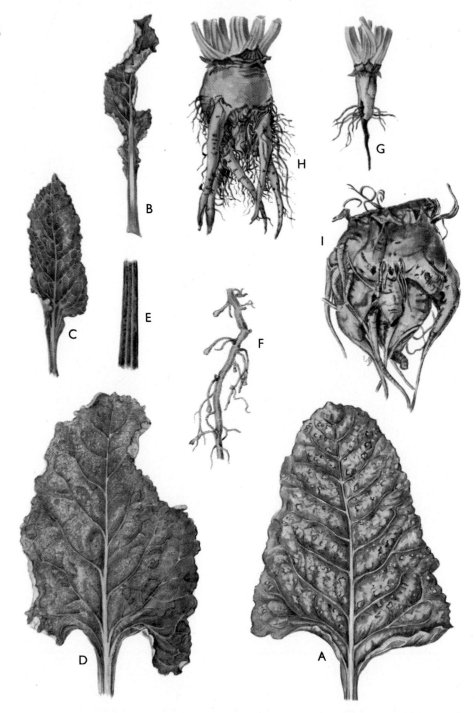

A   Manganese Deficiency (Speckled Yellow) in mangel. Compare symptoms in detail in Plate 57 E-F.
B-F   Chlorate injury on beet plants: B-D leaves in different stages showing yellow speckled discoloration (somewhat like Beet Mosaic: compare Plate 57 E-F) and partly rolled leaf edges; E petiole with necrosis and F root of mangel, rootlets swollen and truncated due to chlorate injury.
G-I   Benzene Hexachloride (HCH) injury in beet. Dry seed treatment with excessive amounts caused root-tip injury in seedlings (G) and resulted in branched, 'tangy' roots (H-I).

Frost injury in beet.
A   Frost necrosis on leaves of a young plant.
B   Mangold seed plant showing frost injury in the leaf tips.
C-D   Frost injury on older beet leaves; the epidermis has partly been loosened and become silvery.
E-G   Frost injury in beet roots. E shows a corresponding normal root; in F severe frost injury in sugar beet (frozen in the clamp), the frozen tissue showing, at right, typical frost cracks; at left, a bacterial decay beginning.

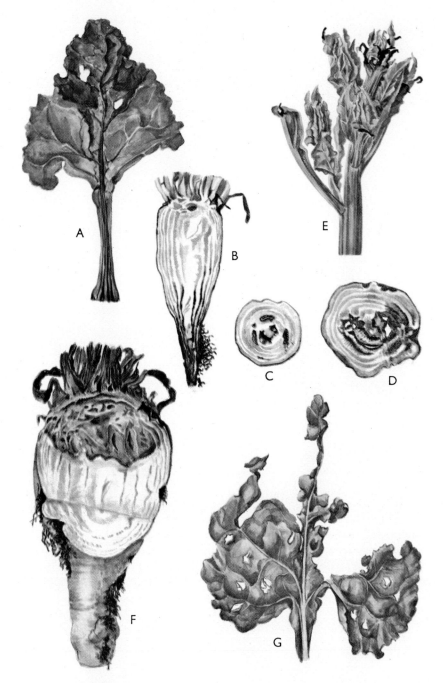

A-D Lightning injury in sugar beet, A in a leaf with scorch symptoms from leaf-tip to base of the leaf-stalk, B longitudinal cracks and C-D concentric cracks in sections of roots.
E Frost injury in the shoot-tips of sugar beet seed plant. May resemble Boron Deficiency. See Plate 50, B-C.
F Frost injury in the upper part of a seed beet growing in the field during winter.
G Sugar beet leaf damaged by hail.

Virus-Yellows *(Beta virus 4)* **in beet** (Compare Plate 56 A-B and Plate 57 C-D).
A   Virus Yellows in beet *(Beta virus 4)*, common symptom.
B   Virus Yellows in beet, speckled type.
C   Virus Yellows, slight apical discoloration and a few fungus spots (Black Mould), caused by *Alternaria spp., Cladosporium,* etc.
D   Virus Yellows in the leaf of a beet hybrid *(Beta maritima* × sugar beet).

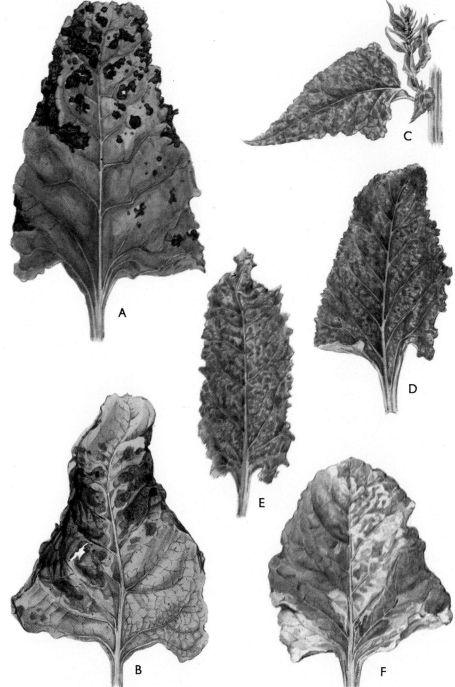

A	Virus Yellows *(Beta virus 4)* in mangold leaf, secondarily invaded by Black Mould *(Alternaria spp.)* and *Phoma betae*.
B	Virus Yellows in sugar beet leaf, almost wholly yellowed and half killed by Black Mould *(Alternaria spp.)*.
C-E	Beet Mosaic *(Beta virus 2)*.
F	Beet leaf with broken colour (albinism caused by mutation).

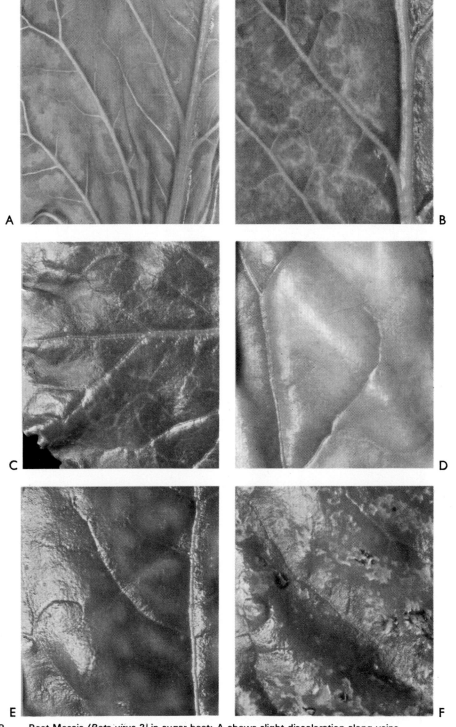

A-B   Beet Mosaic *(Beta virus 2)* in sugar beet: A shows slight discoloration along veins of a heart leaf; B shows typical mosaic discoloration on older leaf.
C-D   Virus Yellows *(Beta virus 4)* in beet leaves: C shows initial discoloration along veins of a young leaf; D typical yellowing on intermediate leaf.
E-F   Manganese Deficiency (speckled yellows) in sugar beet, slight and severe symptoms respectively. Compare Plate 52 A.

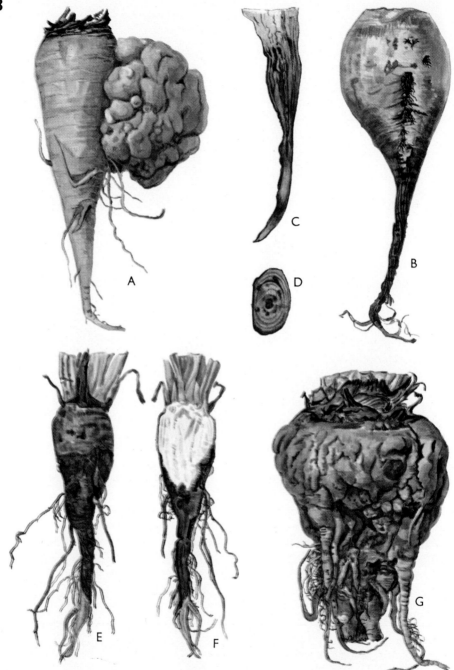

A   Crown Gall *(Agrobacterium tumefaciens)* on sugar beet.
B-D   Tail-rot (probably a bacterial disease) in sugar beet: C a longitudinal section; D a transverse section of the diseased root.
E-G   A Root-rot disease of unknown origin in sugar beet, (probably combination of growth conditions and fungus attack). E-F at midsummer time, G at harvest time.

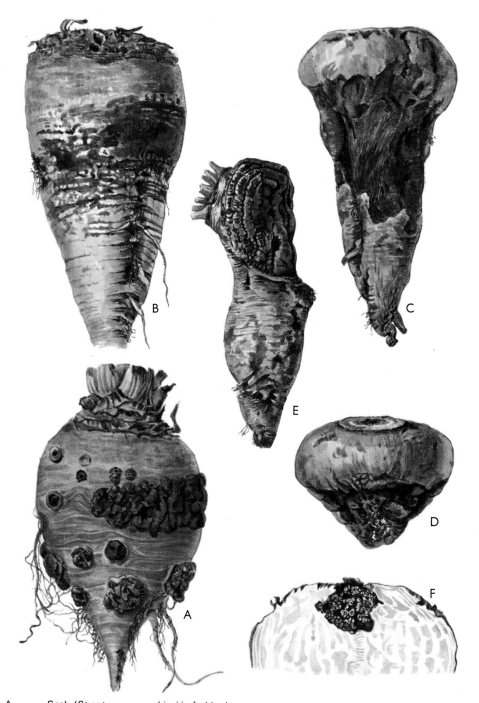

A	Scab *(Streptomyces scabies)* in fodder beet.
B-D	Girdle Scab *(Streptomyces sp.)* in sugar beet, progressing stages from slightly to strongly attacked roots.
E	Sugar beet with mark of early mechanical injury.
F	Cavity normally found in the upper part of big sugar beets, small protuberances formed at the wall of the cavity.

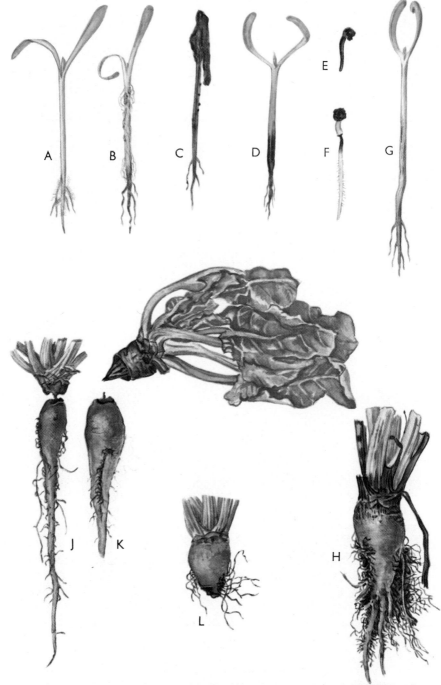

A-G Root Rot (Black Leg) on beet seedlings caused by different Black Leg fungi (C, D and G by *Phoma betae,* E and F by *Pythium debaryanum,* B invaded by *Corticium solani).* A is a healthy seedling.
H Fodder beet (September), abnormally small after chronic Root Rot.
I-K 'Strangles' in mangold (partly due to disturbance by wind [squall] after singling of the seedlings). The root is constricted at soil surface.
L Mechanical damage to mangel.

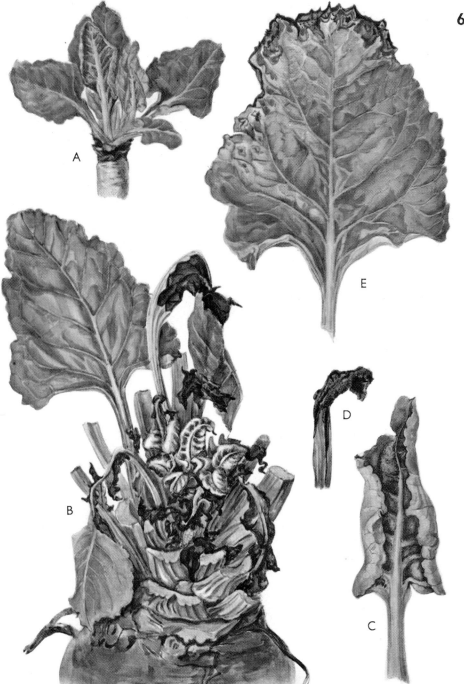

Downy Mildew *(Peronospora farinosa F. sp. betae)* **of Sugar Beet.**
A   Attack in heart of young seed beet plant.
B   Sugar beet, initial attack on the heart-leaves, advanced attack on the older dead leaves.
C   Rolled leaf with fresh conidiophores.
D   Destroyed heart-leaf with dried-up conidiophores.
E   Yellow outer leaf from an affected plant.

A-E   Dark vascular bundles (D-E) in beet, probably due to attack of *Pythium* on the lateral roots. A-C different stages of symptoms in beet-leaves from roots showing attack.
F   Leaf Spots on sugar beet, caused by *Phoma betae*.

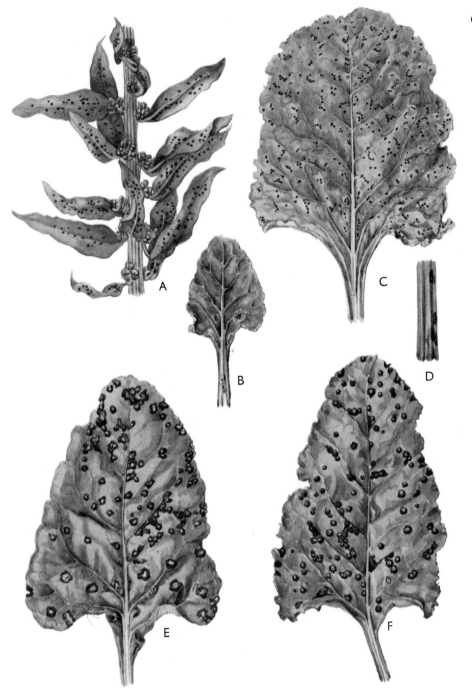

A-D Beet Rust *(Uromyces betae)*: A and C uredo pustules on beet seed plant and sugar beet leaf respectively; B aecidia on leaf of overwintered seed plant; D teleuto stage on sugar beet leaf stalk.
E Leaf Spots *(Ramularia betae)* on mangold leaf.
F Leaf Spots *(Cercospora beticola)* on sugar beet leaf.

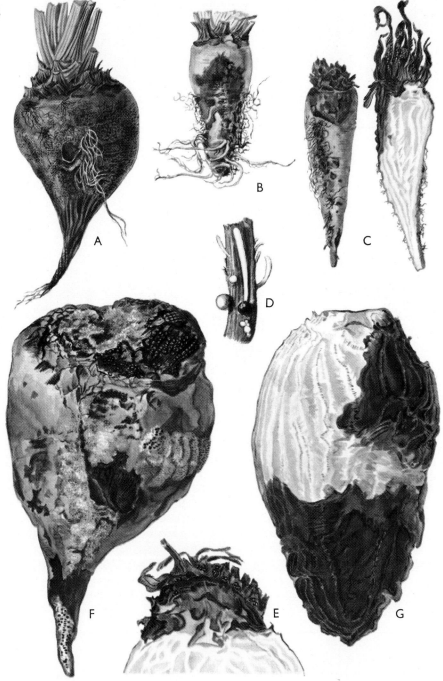

A-B   Violet Root Rot *(Helicobasidium purpureum)* on mangel and sugar beet.
C-E   Root Rot caused by the fungus *Typhula brassicae*: C and E attack in the upper part of seed beets, Sclerotia, whitish, yellow or black, spherical, are seen on the dead tissue; D fruiting bodies (pilei) on dead stem.
F-G   Grey Mould *(Botrytis cinerea)* on beet: F the fungus on the surface of beet from clamp; G section of damaged beet from clamp.

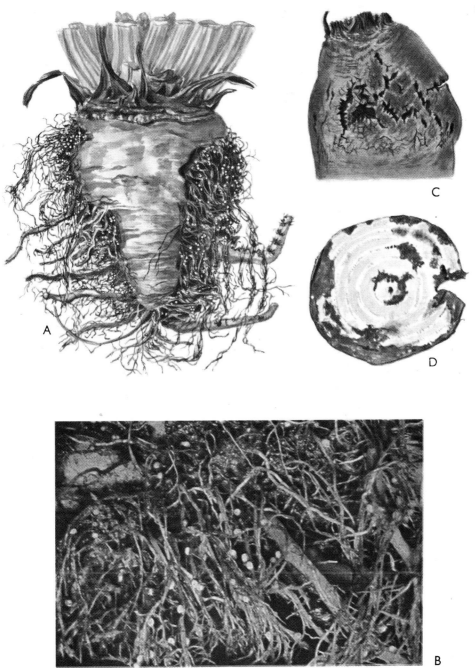

A	Sugar beet with abnormal and excessive development of lateral rootlets due to attack by Sugar Beet Eelworm *(Heterodera schachtii).*
B	Whitish cysts of Sugar Beet Eelworm on rootlets of sugar beet (× 3 mag.)
C	Sugar beet damaged by Stem and Bulb Eelworm *(Ditylenchus dipsaci)*
D	Cross section of sugar beet with Stem and Bulb Eelworm injury.

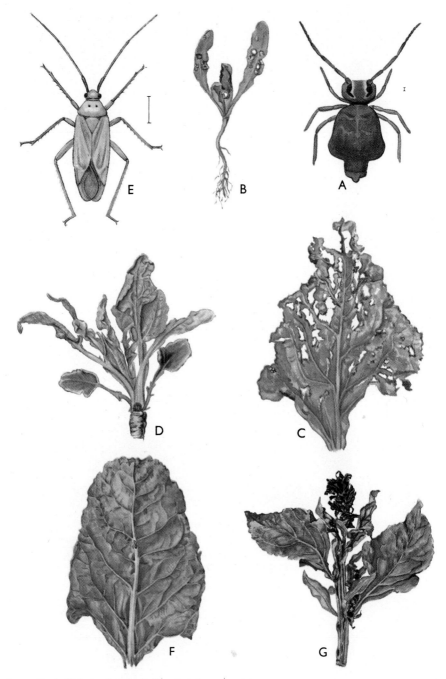

A-B Black Globular Springtail *(Bourletiella hortensis)* and sugar beet plant gnawed by springtails.
C   Leaf of mangold gnawed by The Common Earwig *(Forficula auricularia,* see Plate 36).
D   Sugar beet plant attacked by the Cabbage Thrips *(Thrips angusticeps,* see Plate 81 C).
E-G The Potato Capsid Bug *(Calocoris norvegicus)* and damaged beet plants: F beet leaf showing ulceration at the main rib due to toxic saliva of the bug; G the same on shoot of a beet seed plant. In both cases the distal plant tissues yellowed and destroyed. Damage is most common near hedges.

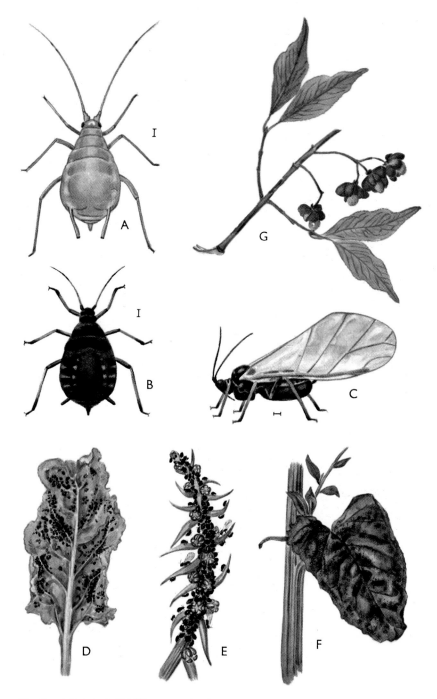

A   The Peach-Potato Aphid *(Myzus persicae)*.
B-C   The Black Bean Aphid *(Aphis fabae)*, wingless and winged forms respectively.
D-E   The Black Bean Aphid on leaf and inflorescence of beet.
F   Sooty Mould (Honeydew fungus) on beet leaf.
G   Twig of Spindle Tree *(Euonymus europaeus),* which is the winter host of the Black Bean Aphid.

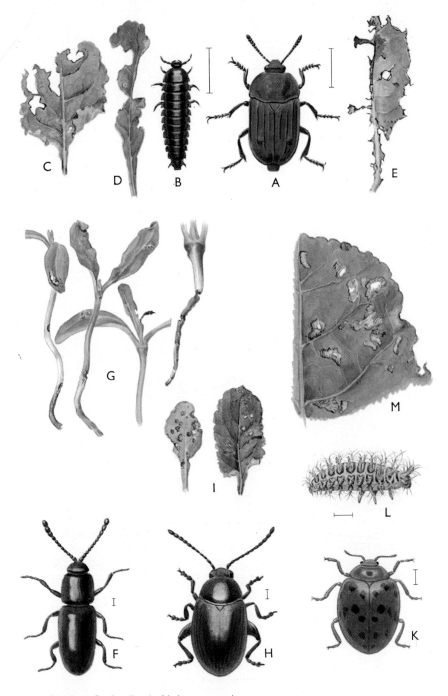

A   The Beet Carrion Beetle *(Aclypea opaca)*.
B   Larva of the Beet Carrion Beetle.
C-D   Sugar beet leaves gnawed by larvae of the Beet Carrion Beetle.
E   Sugar beet leaf gnawed by the Beet Carrion Beetle.
F-G   The Pygmy Mangold Beetle *(Atomaria linearis)* and beet plants injured by the beetles.
H-I   The Beet Flea Beetle *(Chaetocnema concinna)* and beet leaves gnawed by the beetle.
K-M   The Twentyfour-spot Ladybird *(Subcoccinella 24-punctata)* beetle and larva, and their characteristic skeletonising injury in strips on beet leaf.

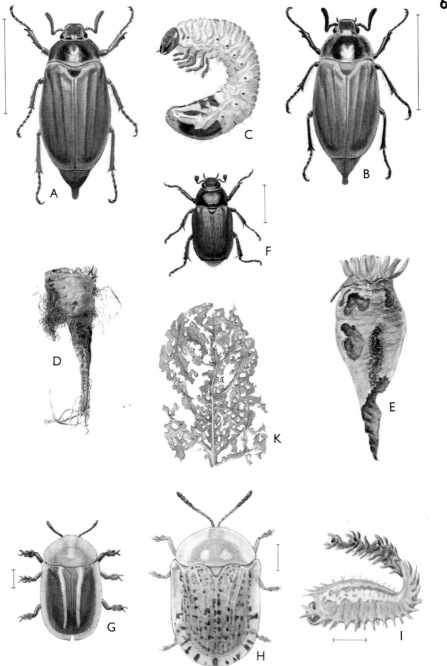

A-B Cockchafers (A *Melolontha melolontha,* B *Melolontha hippocastani*).
C Cockchafer larva.
D-E Beets injured by Cockchafer larvae.
F The Garden Chafer *(Phyllopertha horticola).*
G The Striped Tortoise Beetle *(Cassida nobilis).*
H-I The Clouded Tortoise Beetle *(Cassida nebulosa)* and larva.
K Leaf of mangold, gnawed by the Clouded Tortoise Beetle.

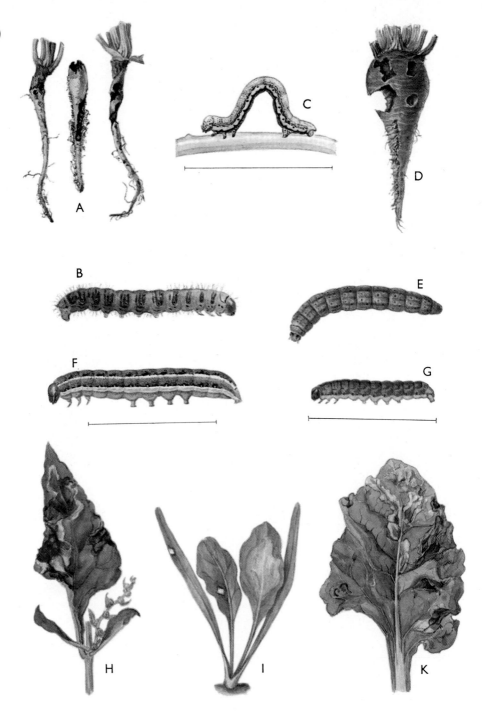

A-B Larva of the Rosy Rustic Moth *(Hydroecia micacea)* and sugar beet plantlets gnawed by the larva.
C   Larva of *Biston zonarius*.
D-E Cutworm *(Agrotis segetum)* and mangold gnawed by the worm.
F   Larva of *Mamestra trifolii*.
G   Larva of the Cabbage Moth *(Mamestra brassicae)*.
H-K Beet leaves mined by the Mangold Fly larvae *(Pegomyia betae)*, I showing eggs on the underside of leaves.

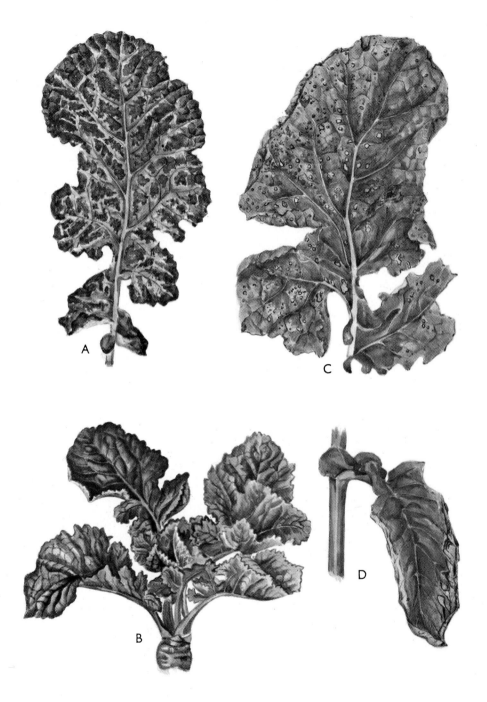

A   Magnesium Deficiency in swede.
B   Phosphorous Deficiency in swede. Compare Plate 91 B.
C-D   Potassium Deficiency in swede.

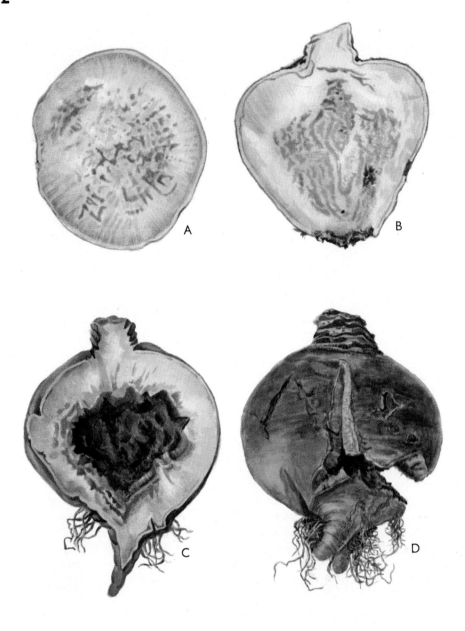

A-C Symptoms of Boron Deficiency (Brown Heart) in swedes, different stages from slight to severe injury.
D   Growth cracks in swedes, probably due to sudden changes between drought and heavy rain and to some extent of hereditary disposition.

A-B Copper Deficiency in swede (B) compared with normal and vigorous plant.
C   Swede seedlings with cold and frost injury symptoms, normal plant at left.
D-E Swede leaves with frost injury symptoms.

A-D Mosaic (probably *Brassica virus 1)* in swedes: A-B in heart-leaves; C-D in seed plants.
E-F Mosaic in turnip.

A   Swede damaged by frost in the clamp, left side of the root collapsed and showing frost cracks.
B   Neck Rot in swede, a secondary symptom of attack of Swede Midge. Compare Plate 88 A and F. In the clamp rot has spread further.
C   Black Rot *(Xanthomonas campestris)* in swede, the root tip mined by Cabbage Maggots, which probably opened the way for the attack.
D   Black Rot on cabbage leaf.

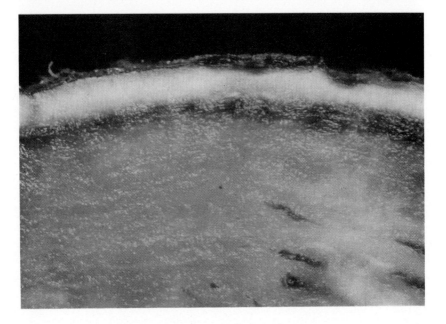

A   Soft Rot *(Pectobacterium carotovorum)* of turnip. To the right of the totally destroyed specimen of healthy plant.
B   Black Rot *(Xanthomonas campestris)* in swede. The cross section (magnified) shows the dark brown vessels.

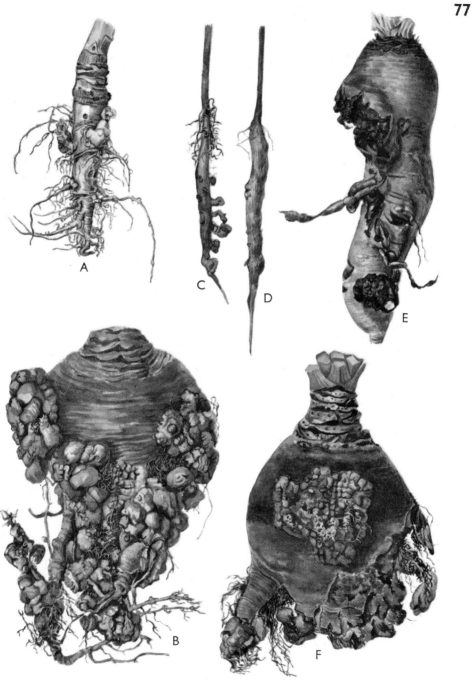

A-B   Hybridisation nodules in swede.
C-D   Club Root *(Plasmodiophora brassicae)* in mustard *(Sinapis alba)*.
E   Club Root on turnip, the older galls partly decayed.
F   Club Root in swede.

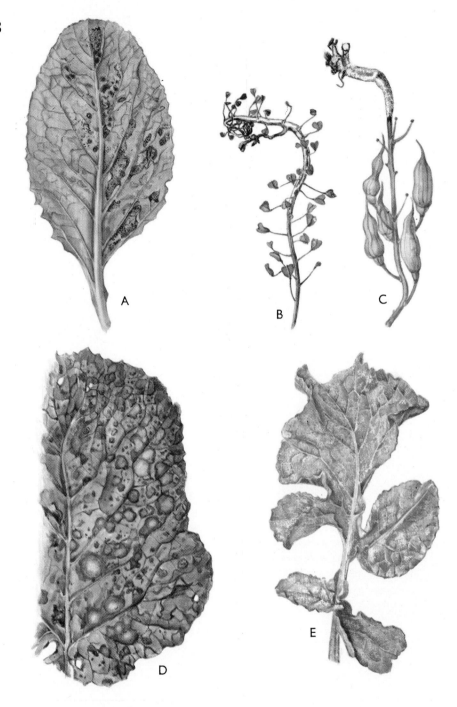

A	Downy Mildew *(Peronospora parasitica)* on cabbage leaf.
B-C	White Blister *(Cystopus candidus)* in shepherd's purse and radish.
D	White Spot in swede, caused by *Cercosporella brassicae*.
E	Mildew *(Erysiphe polygoni)* on swede leaf.

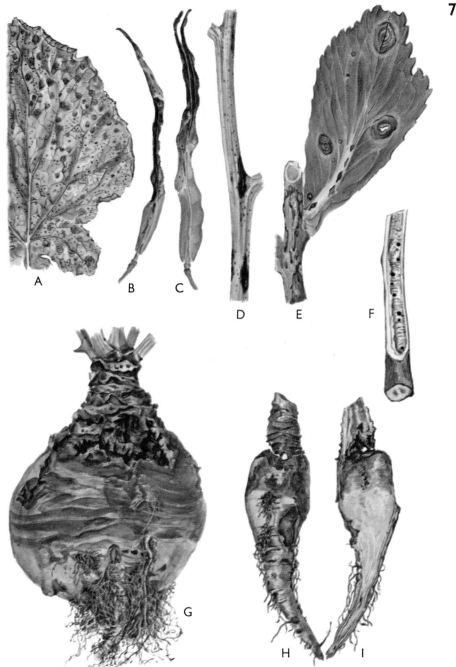

A	Dark Spot *(Alternaria brassicae)* on turnip leaf.
B-E	Dark Leaf Spot *(Alternaria brassicicola)* on pods of cauliflower (B,C), on stem of radish (D) and on leaf and stem of cabbage (E).
F	Spherical, dark sclerotia of the fungus *Typhula brassicae* in the stem cavity of swede seed plant. Compare Plate 64 C-E.
G	Dry Rot Canker *(Phoma lingam)* on swede.
H-I	Dry Rot Canker on swede seed plant.

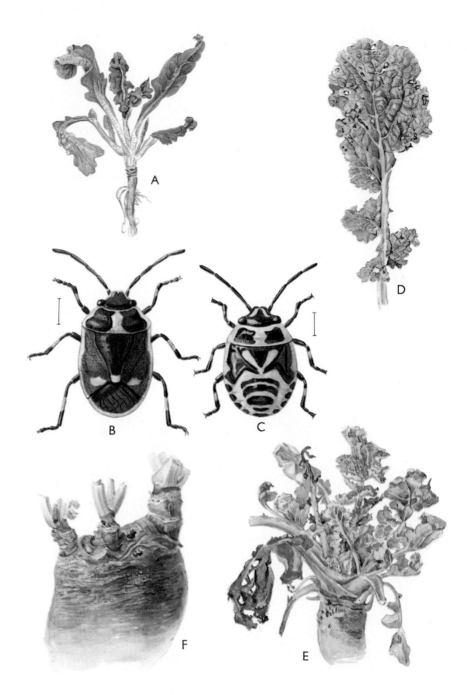

A   Rape deformed by Stem and Bulb Eelworm *(Ditylenchus dipsaci)*.
B-C   Adult and nymph of the Pentatomid *Eurydema oleracea*.
D   Swede leaf injured by *Eurydema oleracea*.
E   Swede injured by *Eurydema oleracea*.
F   Swede with three necks ('many-neck'), the original heart being killed by *Calocoris norvegicus* (the Potato Capsid Bug, see Plate 66 E).

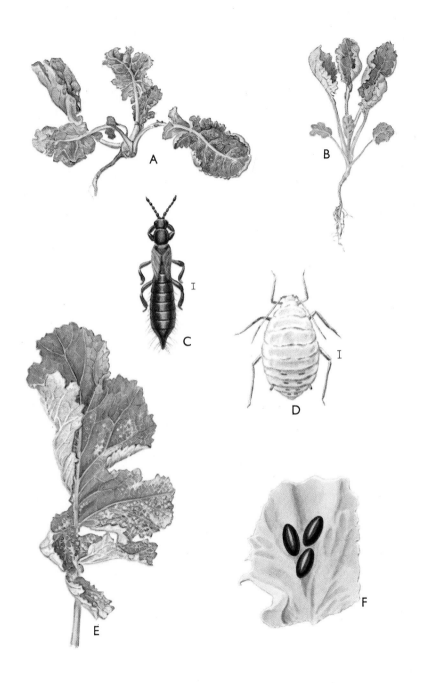

A-B Young swede plants injured by Cabbage Thrips.
C   The Cabbage Thrips *(Thrips angusticeps)*.
D   The Cabbage Aphid *(Brevicoryne brassicae),* wingless female.
E   Swede leaf discoloured and malformed by Cabbage Aphid.
F   Eggs of Cabbage Aphid (highly magnified).

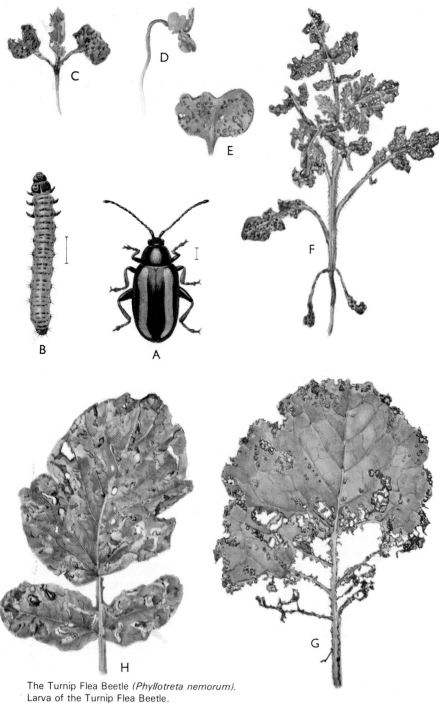

A   The Turnip Flea Beetle *(Phyllotreta nemorum)*.
B   Larva of the Turnip Flea Beetle.
C-D   Flea Beetle injury on swede seedlings.
E   Radish leaf gnawed by Flea Beetles.
F   Flea Beetle injury on mustard plant.
G   Flea Beetle attack on older leaf of swede.
H   Radish leaf mined by numerous larvae of the Turnip Flea Beetle.

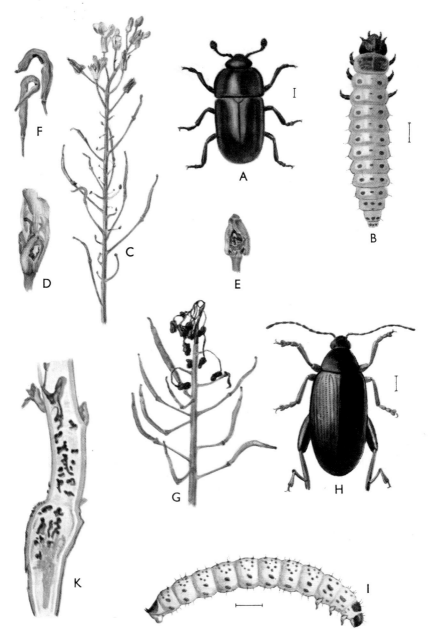

A	The Blossom Beetle *(Meligethes aeneus)*.
B	Larva of the Blossom Beetle.
C	Blossom Beetle injury on swede seed sprout.
D-E	Swede flower buds damaged by Blossom Beetles.
F	Swede pods injured and deformed by Blossom Beetle larvae.
G	Tip of swede seed sprout damaged by larvae of the Blossom Beetle.
H	The Cabbage Stem Flea Beetle *(Psylliodes chrysocephala)*.
I	Larva of the Cabbage Stem Flea Beetle.
K	Attack of Stem Flea Beetle larvae in root and stem of swede seed plant.

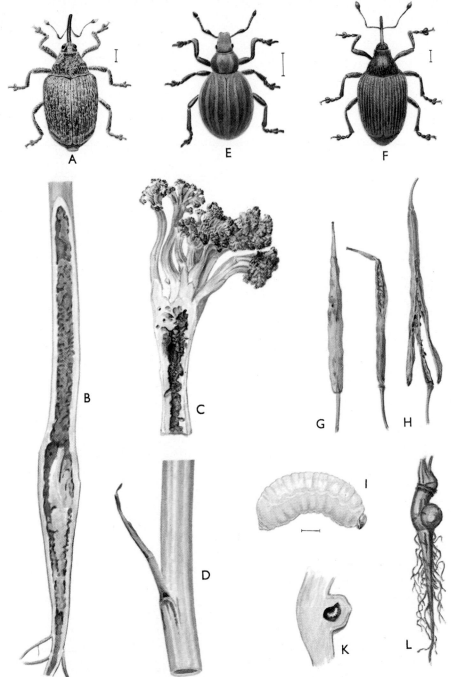

A	The Cabbage Stem Weevil *(Ceutorrhynchus quadridens)*.
B-C	Stems of turnip (seed plant) and cauliflower hollowed out by Cabbage Stem Weevil larvae.
D	Seed stalk of turnip, showing below the petiole a hole where the Stem Weevil larvae bore out.
E	The Sand Weevil *(Philopedon plagiatus)*.
F	The Turnip Seed Weevil *(Ceutorrhynchus assimilis)*.
G	Seed-pod of swede punctured by the Turnip Seed Weevil.
H	Two seed pods of swede damaged by larvae of the Turnip Pod Midge *(Dasyneura brassicae)*.
I	Larva of the Cabbage Gall Weevil *(Ceutorrhynchus pleurostigma)*.
K-L	Galls in swede caused by larvae of the Gall Weevil, K showing larva enclosed in the gall cavity.

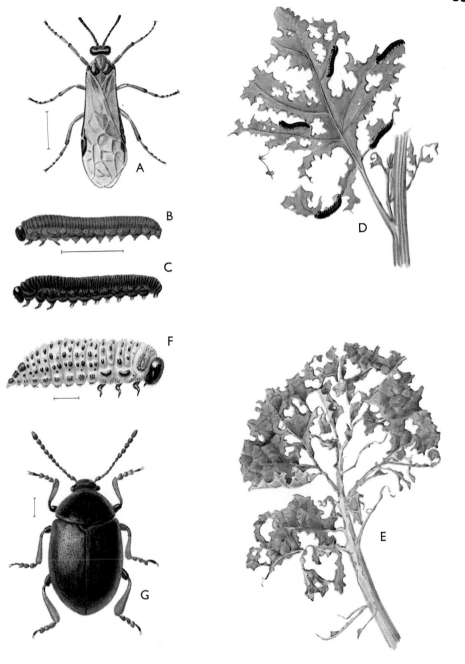

A   The Turnip Sawfly *(Athalia rosae)*.
B-C   Turnip Sawfly larvae.
D-E   Leaves of white mustard and swede gnawed by Turnip Sawfly larvae.
F-G   Imago and larva of the Chrysomelid Beetle *(Colaphellus sophiae)*.

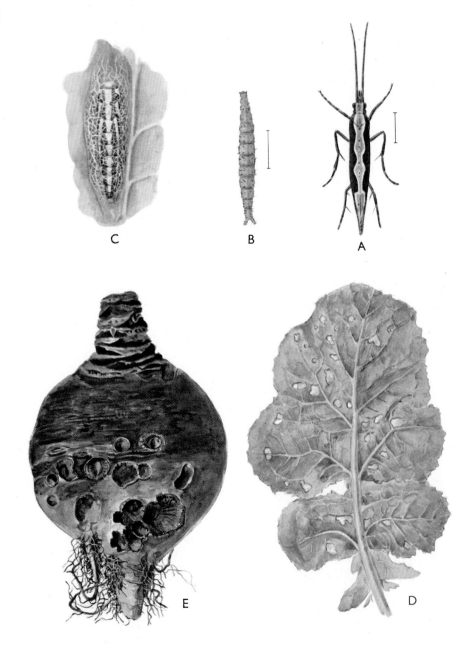

A The Diamond Back Moth *(Plutella xylostella)*.
B Larva of the Diamond Back Moth.
C Pupa of the Diamond Back Moth (in silken cocoon).
D Swede leaf attacked by larvae of the Diamond Back Moth.
E Swede with scars from Cutworm *(Agrotis segetum)* attack.

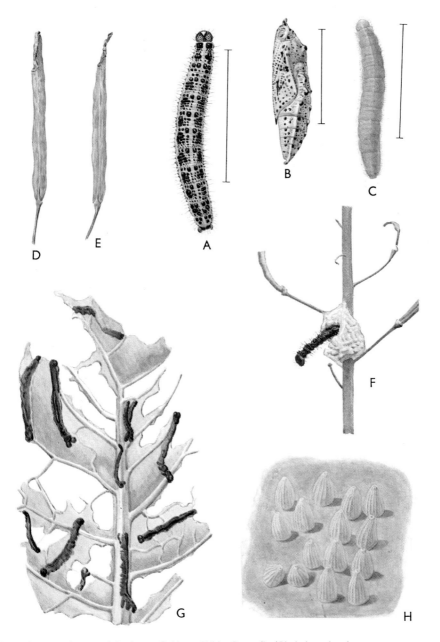

A-B	Larva and pupa of the Large Cabbage White Butterfly *(Pieris brassicae)*.
C	Larva of the Small Cabbage White Butterfly *(Pieris rapae)*.
D-E	Pods of swede gnawed by larva of the Large Cabbage White Butterfly.
F	Large Cabbage White Butterfly caterpillar killed by larvae of the Hymenopterous parasite *Apanteles glomeratus*.
G	Large Cabbage White Butterfly caterpillars on a gnawed leaf killed by parasitic fungus.
H	Egg-batch of the Large Cabbage White Butterfly (magnified).

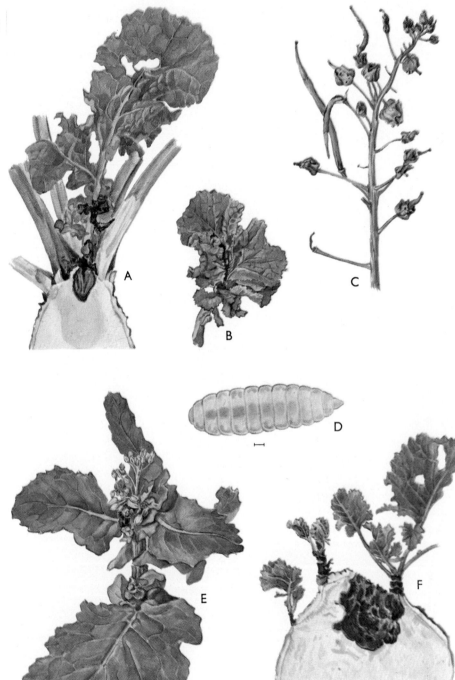

Attack by the Swede Midge *(Contarinia nasturtii)* **on swede.**
A   Attack in the growing point followed by heart-rot.
B   Leaf of swede with attack.
C   Attack by Swede Midge larvae on swede seed plant, showing deformation of flowers.
D   Gall Midge larva.
E   Attack by Swede Midge larvae on swede seed plant, showing deformation of inflorescence.
F   Older attack with dried-up heart-rot and 'many-neck' after larvae feeding in the growing point.

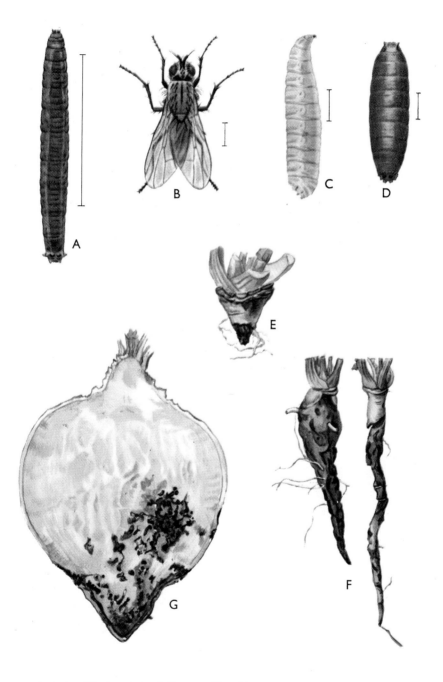

A	Leatherjacket *(Tipula paludosa)*. Compare Plate 24.
B	The Turnip Root Fly *(Erioischia floralis)*.
C-D	Larva and pupa of the Cabbage Root Fly *(Erioischia brassicae)*.
E-F	Attack on swedes by the Cabbage Root Fly maggots.
G	Attack on swede by the Turnip Root Fly maggots.

A-B Potassium Deficiency, variety July, on July 22nd and August 2nd respectively.
C   Injury from storm and drifting sand.
D   Phosphate Deficiency (after W. Krüger & G. Wimmer). Compare Plate 91 A.
E   Injury from storm.

A   Phosphate Deficiency in potatoes grown on reclaimed Calluna heath, ploughed up and marled in 1939. The plot in the foreground has never received phosphate fertiliser, the plot in the background was given 1000 kg superphosphate per hectare in 1940. Photographed in July 1947. For symptoms compare Plate 90 D.

B   Phosphate Deficiency in swedes, same experiment as above, photographed July 1944. For symptoms compare Plate 71 B.

A Manganese Deficiency.
B Magnesium Deficiency. The two leaflets show early and late symptoms respectively.
C Leaf injured by late night-frost.
D Nitrogen Deficiency.
E Borax injury (Borax given erroneously instead of nitrate).

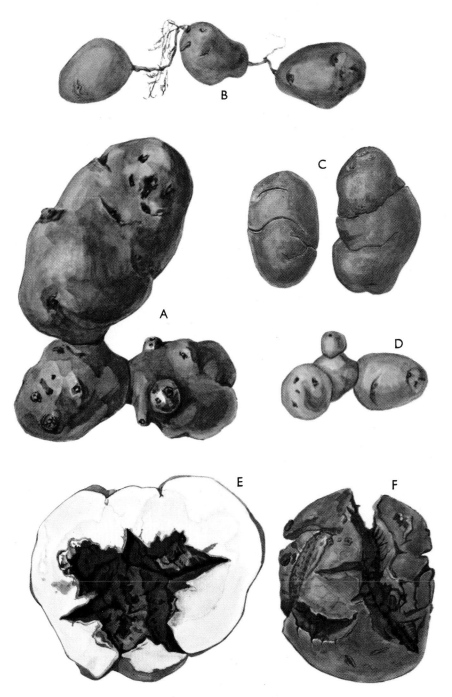

A	Second growth in large tuber.
B	Second growth (pearlstring).
C	'Thumb-nails' and larger cracks of similar character.
D	Second growth in small tuber.
E-F	Growth cracks in very large tuber, following drought and subsequent rain.

A	Net Necrosis (first year infection of Leaf Roll).
B	Internal Rust Spot (probably due to virus).
C	Spraing (probably due to virus).
D-I	Frost injury of varying intensity and type.
J	Tuber proliferation after planting.
K	Spindling sprouts.

A Normal leaf of variety Bintje.
B Rugose Mosaic *(virus Y)* in Bintje.
C Rugose Mosaic *(virus Y)* in King Edward.
D Aucuba Mosaic *(virus G)* in the variety July.

A   Leaf Roll *(Solanum virus 14)* in Majestic, infected the same summer.
B   Leaf Roll in Birgitta, infection from previous years.
C   Leaf Roll in King Edward.
D   Rolling leaves in King Edward, caused by drought and too late earthing up.

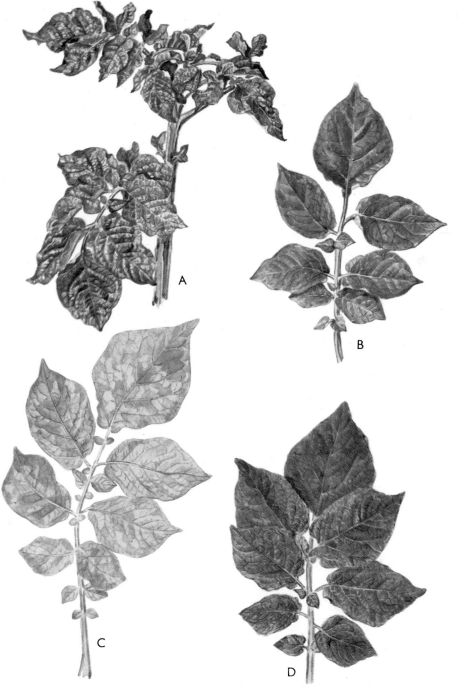

A	Crinkle *(virus A + X or Y)* in the variety Frühgold.
B	Healthy leaf of Frühgold.
C	Mild Mosaic *(virus X)* in the variety Bintje.
D	Mosaic and dark colour *(virus S)* in seed plant.

A   Leaf Drop Streak *(virus Y)*.
B   Underside of leaflet with severe streak symptoms.

A   Black Leg *(Pectobacterium carotovorum var. atrosepticum)*.
B   Black Leg in young tuber infected from stolon.
C   Section of stem in which the pith is rotted away at the base.

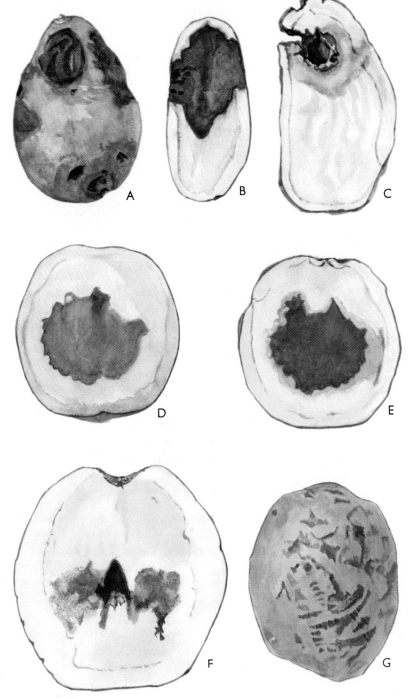

A-B Bacterial soft rot *(Pectobacterium carotovorum)* or other bacteria.
C   Rot due to mechanical injury during lifting.
D-F Types of heating injury (variety Tylstrup Odin).
G   Colonies of *Bacterium prodigiosum* on cooked and peeled potato.

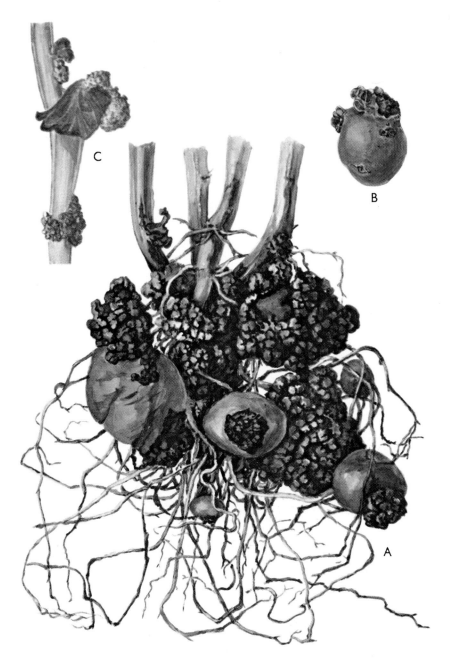

A   Wart Disease *(Synchytrium endobioticum)*.
B   Powdery Scab *(Spongospora subterranea)* on tuber with second growth.
C   Wart Disease on basal leaves.

A-B Blight *(Phytophthora infestans)* on upper and lower leaf surface respectively.
C   Blight on stem (month of July).
D-F Leaf spots caused by *Cercospora concors:* D early symptoms on upper leaf surface; F late symptoms; E conidia on lower surface of leaflet.
G   Early Blight *(Alternaria solani)*.

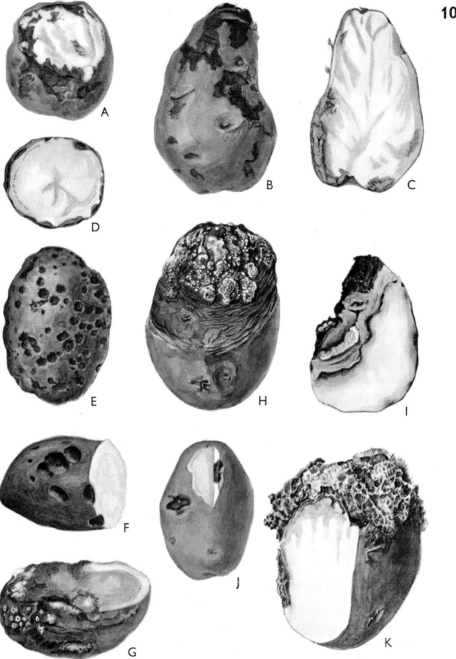

A	Late Blight *(Phytophthora infestans)* in tuber.
B-C	Mechanical injury during lifting, showing bruises and patches of starch under the skin.
D-E	Pit Rot, type with small spots.
F	Pit Rot, large spots.
G	Dry Rot due to *Fusarium coeruleum*.
H-I	Dry Rot due to *Fusarium roseum*.
J	Lesions on tuber due to the Early Blight Fungus *(Alternaria solani)*.
K	Dry Rot due to *Fusarium sulphureum*.

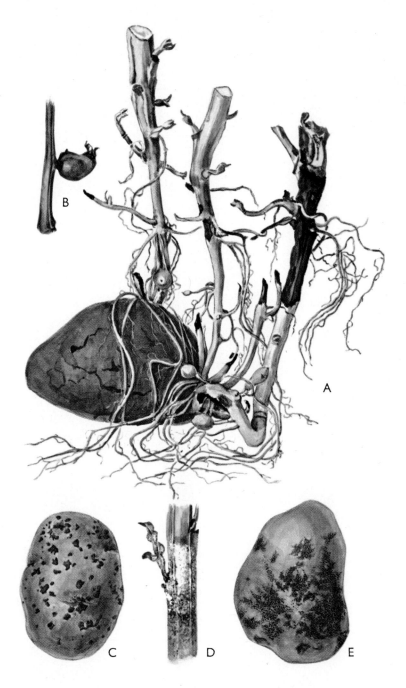

A	Attack of *Corticium solani* on potato sprouts, stems and stolons.
B	Aerial tuber, sequel to injury on potato stem bases, due to *Corticium*, cutworms, etc.
C	Black Scurf, the resting stage of *Corticium solani*.
D	*Thanatephorus cucumeris*, the basidial stage.
E	Violet Root Rot *(Helicobasidium purpureum)*, microsclerotia on potato.

A-B	Common Scab *(Streptomyces scabies)*, slight and severe.
C	Powdery Scab *(Spongospora subterranea)*, galls on potato root.
D	Powdery Scab.
E	Silver Scurf *(Spondylocladium atrovirens)*.
F	Skin Spot *(Oospora pustulans)*.
G	Russeting, due to fluctuating rainfall and varietal effect.
H	Incrustations due to combined attack of scab, mites, various maggots, and cracks.
I	Feathered skin on second growth.

A   Potato Root Eelworm *(Heterodera rostochiensis)*. Plant with weak haulm and abnormally branched roots with numerous cysts.
B   Cysts of Potato Root Eelworm, the younger ones white, the older yellow or brown (greatly magnified).
C   Potato attacked by the Potato Tuber Eelworm *(Ditylenchus destructor)*.

A Colorado Beetles *(Leptinotarsa decemlineata)* and larvae on potato haulm.
B Colorado Beetle.
C Colorado Beetle larva.

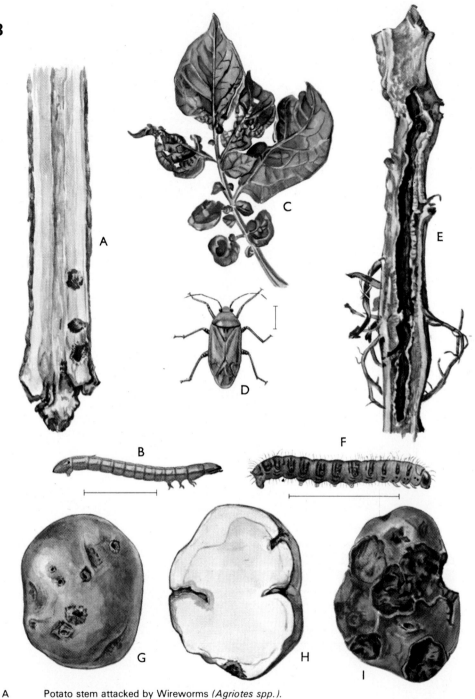

A	Potato stem attacked by Wireworms *(Agriotes spp.)*.
B	Wireworm *(Agriotes sp.)*.
C	Potato leaf injured by Common Green Capsid *(Lygocoris pabulinus)*.
D	Common Green Capsid *(Lygocoris pabulinus)*.
E	Potato stem attacked by the larva of the Rosy Rustic Moth *(Hydroecia micacea.*
F	Larva of the Rosy Rustic Moth *(Hydroecia micacea)*.
G	Potato damaged by Wireworms, corky scales around the holes.
H	Potato damaged by Wireworms, the lower lesion invaded by *Corticium solani*.
I	Tuber gnawed by larvae of Cockchafers *(Melolontha melolontha)*.

A	*Sclerotinia sclerotiorum,* sclerotia and fruiting bodies *(apothecia).*
B	*Sclerotinia sclerotiorum* on carrot in storage.
C	Black Rot *(Stemphylium radicinum)* and other fungi on carrot from pit.
D-E	Frost injury with distinct cracks.
F	Millipede *(Blaniulus guttulatus).*
G	Millipede injury.
H	Violet Root Rot *(Helicobasidium purpureum)* on carrot.

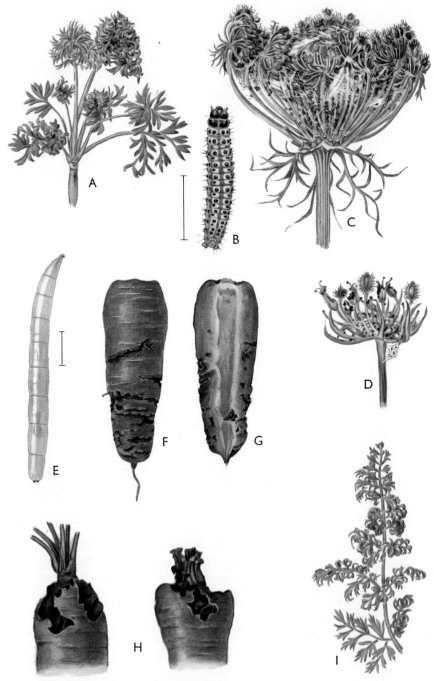

A	Carrot top with attack of *Trioza apicalis*.
B	Larva of Seed Moth *(Depressaria sp.)*.
C-D	Umbels of carrot with injury and web from Seed Moths.
E	Larva of Carrot Fly *(Psila rosae)*.
F-G	Carrots injured by Carrot Fly larvae.
H	Injury from Cutworms *(Agrotis sp.)*.
I	Carrot leaf attacked by aphids.

A   Manganese Deficiency in flax.
B-C   Flax Seedling Blight *(Colletotrichum lini)*.
D-F   Browning and Stem-Break *(Polyspora lini)*.

A-B Flax Rust *(Melampsora lini)* summer spores and winter spores, respectively.
C-D Pasmo *(Septoria linicola)*, pycnidia and leafspots, respectively.

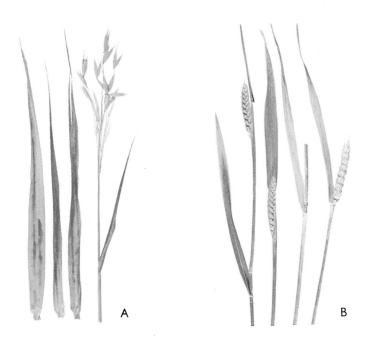

A   Oats with retarded root growth (Barley Yellow Dwarf Virus).
B   Wheat plants with Barley Yellow Dwarf Virus.
C   Oats with retarded root growth, at the edge of an oatfield.

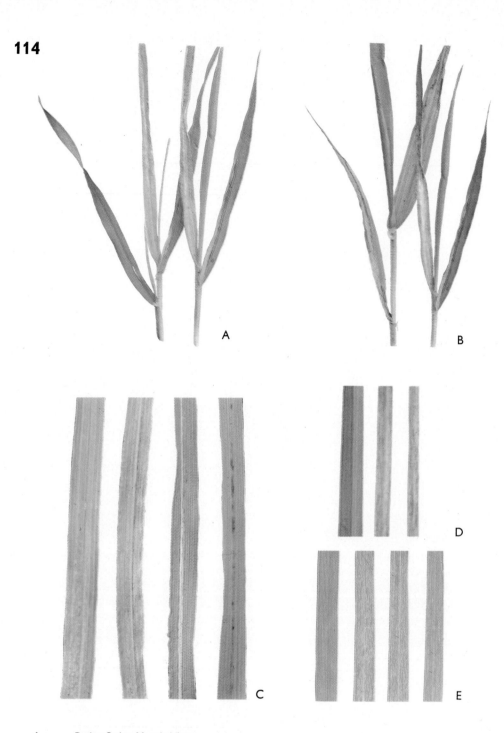

A	Barley Stripe Mosaic Virus.
B	Leaf Stripe of barley *(Pyrenophora graminea)* on the left, compared with Barley Stripe Mosaic Virus on the right.
C	Barley Stripe Mosaic Virus on barley leaves.
D	Bearded Couch Grass Mosaic Virus, on the left a healthy leaf.
E	Rye Grass Mosaic Virus, on the left a healthy leaf.

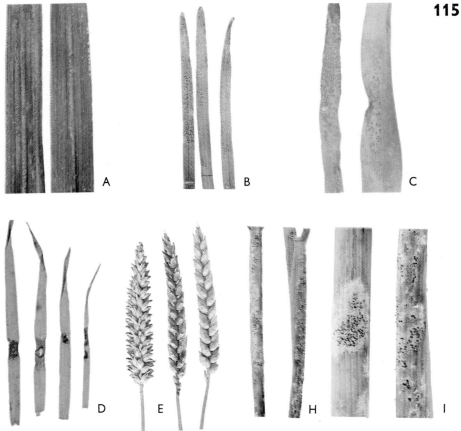

A-C	Yellow Rust of Wheat *(Puccinia striiformis)* with pustules in stripes on older leaves (A) and without stripes on leaves of young wheatplants.
D	Glume Blotch of Wheat *(Septoria nodorum)* on young wheat leaves.
E	Glume Blotch of Wheat on young ears of wheat, on the right a healthy ear of wheat.
F-G	Glume Blotch of Wheat on older ears of wheat, note in F the flask-shaped bodies containing spores (pycnidia) on the glumes, greatly magnified.
H-I	Mildew *(Erysiphe graminis)* with mycelium and minute, spherical fungal fruiting bodies containing spores (cleistocarp) on older barley plants, I greatly magnified.

A-C   Grain Aphids *(Macrosiphum avenae)* on a barley leaf and in an ear of wheat with respectively green and reddish-brown individuals. In B two greatly swollen, parasitised aphids can be seen.
D-E   Rose Grain Aphids *(Metopolophium dirhodum)* in colonies with respectively a female without wings and one with wings.
F-G   Bird Cherry Aphids *(Rhopalosiphum padi)* on respectively a leaf and lower stem of barley.

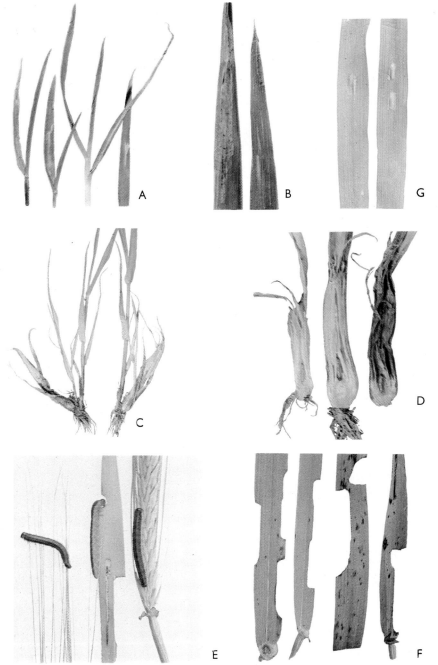

A-B	Wheat Leaf Mining Fly larvae *(Hydrellia griseola)* in respectively volunteer plants from dropped barley seed (winter barley) and in older leaves of wheat. See Plate 26 E.
C-D	Gout Fly larva *(Chlorops pumilionis)* on the main shoot of winter barley, in D the larvae can be seen (magnified).
E-F	Saw Fly larvae *(Dolerus sp.)* on ears of barley and badly bitten leaves.
G	Saw Fly eggs placed in pockets on young barley leaves.

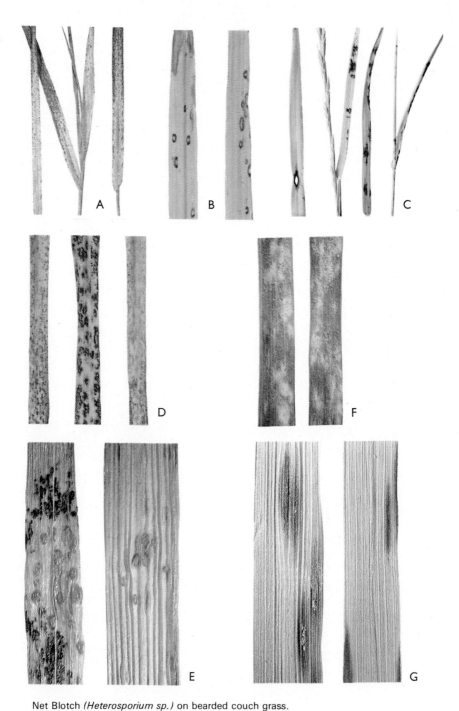

A	Net Blotch *(Heterosporium sp.)* on bearded couch grass.
B	Leaf Streak *(Heterosporium phlei)* on timothy grass.
C	Leaf Blotch *(Helminthosporium sp.)* on Italian rye grass.
D-E	Crown Rust *(Puccinia coronata)* on Italian rye grass, note in E the black teliospores around the uredospores, greatly magnified.
F	Mildew *(Erysiphe graminis)* on meadow grass.
G	Leaf Spot *(Mastigosporium rubricosum)* on bearded couch grass (greatly magnified).

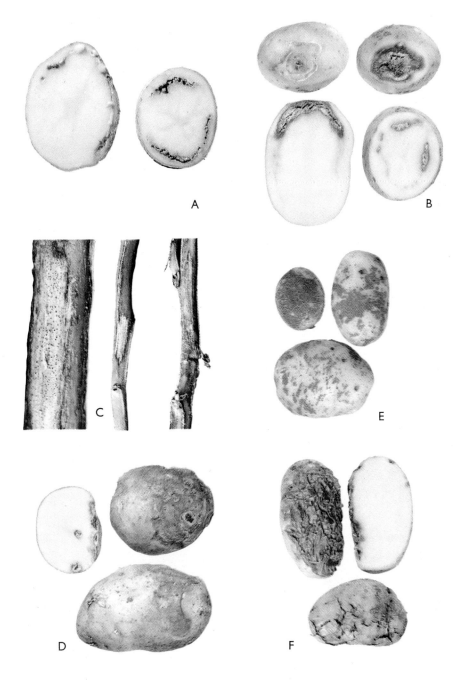

A	Bacterial Ring Spot of potatoes *(Corynebacterium sepedonicum)* on the variety Record.
B	Damaged potato tubers due to spraying with Reglone (diquatdibromide).
C-D	Diseased stalks and potato tubers (gangrene), caused by the fungus *Phoma exigua*. Note on the stalk to the left the dark flask-shaped bodies containing spores (pycnidia).
E	Common Scab *(Streptomyces sp.)* on the variety Bintje.
F	Potato Tuber Eelworm *(Ditylenchus destructor)* in potato tubers.

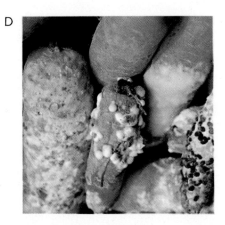

A   Root Knot Eelworm *(Meloidogyne hapla)* in carrots.
B   Boron Deficiency symptoms on the leaves of carrots.
C   Cavity Spot of carrots.
D   Common Grey Mould *(Botrytis cinerea)* or Bacterial Soft Rot *(Erwinia carotovora)* in stored carrots.
E-F   Crater Rot *(Corticium carotae)* of stored carrots, in F infection from the wooden box can be seen.

Introduction

Unlike most works on pests and diseases in agriculture which deal with only one aspect of the subject, this book gives serious consideration to all of the three major areas involved: the colour plates enable rapid identification; there are notes on life cycles, symptoms and damage; and up-to-date control measures for individual disorders are described.

The text explains the basic structure and classification of pests and diseases and goes on to give an insight into the causes of physiological disorders. This is extremely important as it is vital to know the exact cause of a problem before attempting to treat it. Without this knowledge control measures can be, and unfortunately in practice often are, an expensive waste of time. It is also important to know how pests and diseases are transferred from one area to another: the reasons can be very diverse, ranging from aphids carrying viruses on their feeding mouthparts to splashes of water carrying bacteria from one leaf to another.

Having identified the disorder with certainty and decided that it is economic to treat it, the grower will have four basic methods of control to choose from: cultural, chemical, biological or an amalgam of the three, called integrated control.

Cultural controls are the oldest form of prevention and restriction of plant disorders. They encompass soil improvements, rotations and the following of correct cultural practices.

Chemical control is exactly what its name implies: some form of chemical is applied either to the soil or the crop to prevent the disorder becoming established, or in some cases to eliminate it before a check in growth has occurred. Many chemicals are now available for the control of pests and diseases, but great care must be taken in using them as they can be very dangerous to plants, animals and human beings.

The use of biological control is rather restricted where climatic conditions are very variable. Notable examples of its use are the control of Red Spider Mite in indoor and outdoor crops using the mite *Phytoseiulus persimilis* and the general control of pests by species such as Ladybirds, Black-kneed Capsids, Lacewings and Ground Beetles. It is not used for the control of pathogenic disorders.

Although the fourth method, that of integrated control, has worked for many years it has only recently been developed into a highly sophisticated scientific study. It brings together the cultural, chemical and biological control measures in such a way as to derive an optimum balance of the benefits of each method.

One other method, not strictly included in the previous classes, which is sure to develop in the coming years is that of breeding strains of plants specially resistant to diseases and pests.

The prevention and control of pests and diseases is a large subject, and the following table has been included to clarify the alternatives which are available to the grower.

Frank Hope 1979

Summary of Methods of Pest and Disease Control

To attack the pest or disease
EXCLUSION
By interception of materials before they enter the country.
By destruction of pests and diseases, using sprays, disinfectants, dusts and fumigants.
By isolation of plant material from infected/infested stock.
By prohibiting the importation of certain seeds and materials.

ERADICATION
By cultivation, i.e. the destruction of over-wintering pests and weed hosts.
By rotation of individual crops.
By removal of pests and diseases by roguing out infected stock.
By destruction of pests and diseases using poisonous chemicals.

To protect the plant host
IMMUNISATION
By nutrition: applying fertilisers can give temporary immunity to diseases.
By plant breeding to produce resistant strains.
By application of chemical sprays to prevent the pest or disease from becoming established.
By inoculating with chemicals to prevent disease.

PROTECTION
By applying protective sprays to plant leaves.
By destruction of dissemination agents.
By cultivations, in that the correct cultural practices can reduce the incidence of attack.
By trapping pest species in specially made traps.

PESTS, DISEASES AND CONTROL

Plant Pests

Insects and other pests exist in vast numbers and attack various parts of agricultural crops, but fortunately for the grower only a small number of them cause significant economic damage. However, the damage which does occur represents a vast amount of money in lost yields and expensive sprays. Thus it is very important that pest species are identified at an early stage. It should also be noted that by the time an untrained eye has recognised a specific symptom or pest species, then a great deal of damage may have been caused. This means that where growers wish to identify troubles early on they should learn the major signs of damage and pest species of their specific crops, so that effective protectant or eradicant measures can be taken.

Unfortunately, plants may have more than one species of pest attacking them at any one time and the grower must look closely to ascertain the major cause of damage. This situation can be complicated by the fact that the pest may have moved to another crop, be nocturnal, or be too small to be seen by the naked eye.

The major pests of farm crops fall into the six distinct groups discussed below.

Insects

Insects are found living in every sort of agricultural situation from below ground to the aerial parts of plants. Some do little visible damage to crops, but effects such as stunting, and the part they play in spreading disease and viruses can lead to large yield reductions. Insects belong to the group – the Phylum – of invertebrate animals known as the Arthropods. They are characterised by having compound eyes, hard exoskeletons and bodies with three distinct parts, the head, thorax and abdomen.

At some stage of their life cycle all insects lay eggs which hatch to produce larvae. Sometimes the larvae differ greatly from their parents in physical form. When this is the case they are called grubs, caterpillars or maggots depending upon the species. The larval stage of the insect life cycle is usually the longest and is the stage where the major feeding and growth takes place. When fully fed, the larva enters a new phase and is called the pupa. At this stage it rests and reorganises itself into the adult, completely different in appearance. When there are four distinct stages, egg, larva, pupa and adult, this is called complete metamorphosis. There are some species which have only three distinct stages of growth namely egg, nymph and adult and these are said to exhibit incomplete metamorphosis. Of insects which undergo incomplete metamorphosis, the nymphs are similar to the adults except in size. As they develop they become larger and go through several 'moults' before they become adults.

Particular insects are controlled at specific stages of their life-cycle though this may not be the stage which does the most damage. There are of course large numbers of insects which are beneficial to farmers due to their habits of either eating or parasitising pest species. The major orders of insects are Collembola, an example of which is the Springtail; Orthoptera, e.g. Grasshoppers, Cockroaches; Dermaptera, e.g. Earwigs; Thysanoptera, e.g. Thrips; Homoptera, e.g. Aphids, Leafhoppers, Mealybugs; Coleoptera, e.g. Beetles and Weevils; Lepidoptera, e.g. Butterflies and Moths; Hymenoptera, e.g. Bees, Wasps, Sawflies; Diptera, e.g. Flies, Gnats.

Mites

Mites belong to the class Arachnida which makes them relatives of the common spider. Arachnida is a group containing numerous species of varying economic importance. Their habitats, too, are wide ranging: many live in plant galls while others feed on the undersurface of leaves, or on grain in stores. The species found as pests of agriculture are usually only just visible to the naked eye. They have no clear distinction between body segments, no antennae, and have six legs when young, and eight when mature.

On outdoor crops their attacks are frequently restricted to the undersurface of leaves where they suck the sap causing a distinct bleached speckling. This can lead to severe stunting of growth, defoliation or even death of a crop. They have the ability to increase in numbers rapidly. In general, the higher the temperature and humidity the faster they develop.

Chemicals known as acaricides have been developed specifically to control mites. Integrated control, however, a combination of chemical with other methods, has become popular, as some species develop resistance to chemicals after persistent usage.

Eelworms

This is an important group of pests containing numerous species which attack agricultural crops. They have a wide host range and species are found attacking roots, stems, bulbs, corms, tubers, leaves and flowers, whilst some are free living in the soil being beneficial or attacking crops as secondary agents.

Eelworms or nematodes are minute creatures that are usually visible only under strong hand lenses or microscopes. They are transparent worm-like creatures with tapering ends and although they can travel short distances in moisture films, their movements are restricted. They are spread mainly by soil movement on tractors, workers' clothing and harvested crops. Once established in a soil their multiplication can be rapid, and this is intensified by insufficient rotation or where populations of susceptible weeds are allowed to grow.

Damage to crops varies with the species, some types causing for example the production of cysts on roots or wilting in sunny weather. Other symptoms include paling of lower leaves, staining of tissues, especially in bulbs, speckling of foliage or dead patches of crops. Because of this diversity, symptoms are described in the text under their specific headings.

Slugs and Snails

Slugs and snails are mainly nocturnal creatures, thriving in moist situations rich in organic matter. They are hermaphrodite and each adult can lay large clusters of oval, transparent eggs. Once hatched, the young live on dead or decaying organic matter, but they soon move to attack seedlings, leaves, stems, roots or tubers. In some circumstances, such as in potato crops, damage is often mistaken for that caused by wireworms. Slugs and snails can feed throughout the year. They feed by pulling their food over a ribbed tongue called a radula.

It has proven very difficult to obtain 100 per cent control of slugs and snails in the field, especially on heavy land during wet seasons. A satisfactory measure of control is however achieved by a combination of natural predators such as birds, insects and moles, with traps and chemical baits.

Millipedes and Centipedes

Millipedes are hard bodied, shiny creatures which are often mistaken for centipedes and very occasionally wireworms. They live in moist soil and move in a sluggish manner. When disturbed they tend to coil up like clocksprings. Their bodies are cylindrical with two pairs of legs per segment, and they range in colour from dark grey-brown to pale yellow.

The main source of food for millipedes is dead or decaying organic matter, but they will attack germinating seeds such as peas, and gnaw the roots of a variety of plants such as potatoes.

Centipedes however are beneficial creatures that live on small slugs, snails and earthworms. They differ from millipedes by being very active, having flattened bodies and one pair of legs per body segment.

Woodlice

Woodlice are more of a problem in gardens and smallholdings than large scale agricultural units. They are nocturnal creatures and can be identified by having seven pairs of legs, segmented, flattened bodies, and two pairs of antennae. They live in moist shaded positions and feed on dead organic matter, but some species do attack outdoor plants.

In addition to the above mentioned pests it is often found that localised but serious damage is caused by birds and wild mammals. Canada geese, bullfinches, rabbits and hares are particularly notorious. As these troubles are often sporadic it is often difficult to effect good control measures, but the use of bird scarers, chemical repellents, shooting and in some cases disease introduction, will usually give adequate control.

Plant Disorders

Plant pathology is the study of any variation from normal growth and it is used to ascertain why a crop has not produced an acceptable yield. Important factors for healthy plant life are discussed below.

The Correct Genetical make-up of Seeds

Poor genetical make-up can lead to uneven growth, variability in maturity, production of 'sibs' such as in Brussels sprouts, the formation of chimaeras, incompatability and albinos. Extensive breeding programmes have succeeded in producing various advantageous characteristics. The development of the F_1 hybrid is arguably the most important of these because it has helped to increase yield, stabilise maturity, produce uniform growth and further the possibility of standardising crops – important if growing for specific markets.

Seed Sowing and Planting

Seeds must be sown at the optimum depth and in a seedbed which will encourage rapid establishment. With the production of F_1 hybrids and the general improvement of seed samples, sowing rates and planting densities have become of paramount importance. Research establishments have carried out experiments on these topics and introduced formulas to maximise yields.

Skilful Crop Management

Skilful management of crops can tip the balance between an acceptable or good yield and a total crop failure. Good crop management is difficult to define, as growing techniques and soils vary. Knowing exactly when to carry out operations on a particular site comes only with years of practice, backed up by research work and discussions with other, especially local, growers.

Correct Environmental Conditions

Plants are very demanding of proper growing conditions. Both growth rate and yield are affected by the right balance of essential nutrients and correct amounts of water, heat, air and light. Growth can otherwise be very erratic.

Mechanical and Chemical Injury

Running tractors through crops during cultivations, thinning beet, over-compaction around young plants, rolling crops in adverse weather or at the wrong stage of growth and untimely hailstorms can to a greater or lesser degree reduce the yield of a crop.

Chemical injury can produce results ranging from the total death of large areas of crops to minor speckling of foliage. Damage occurs principally in the following ways: the over-application of chemicals by severely overlapping, faulty nozzles or applying at too high a dosage; the use of incorrect materials; applying chemicals in 'cocktails' where one of the constituents is incompatible with the others, and, as commonly occurs, damage caused by insufficient cleansing of sprayers after use.

Plants require twelve essential nutrients: some, the major elements, are required in large quantities and others, the minor elements, are needed in minute amounts. Of the major nutrients, nitrogen is used mainly for the production of proteins such as chlorophyll; phosphorus is used in the formation of nucleo-proteins and energy transfer; potassium is used in the process of osmotic pressure; calcium is used in plant metabolism and structure; magnesium is a constituent of chlorophyll, and sulphur is used in amino acids.

Of the minor nutrients, iron is used as a catalyst in the production of chlorophyll; manganese, copper and zinc are used in the activation of enzymes; boron is used in the movement of plant sugars and molybdenum is essential for nitrogen producing enzymes.

Although toxicity problems and deficiencies of minor nutrients are rare, when a deficiency does occur the symptoms are serious. Deficiencies of the major nutrients are more common.

Infectious Diseases

Vast amounts of chemical sprays are applied to alleviate the serious reductions in yield caused by infectious diseases. The grower must first recognise the symptoms and then select a system of protection or eradication. The problem in the initial identification is that a range of causes can sometimes produce the same symptoms. Where a diagnosis is uncertain it is advisable to contact the local A.D.A.S. advisor who will give a positive identification and recommend a safe and efficient control.

The major infectious diseases which are found affecting plant growth are caused by fungi, bacteria, actinomycetes, viruses and mycoplasmas.

FUNGI. The main body of a fungus usually consists of a number of branching threads called *hyphae*, collectively termed *mycelium*. The hyphae are only one cell in thickness but often have numerous cross walls, called *septa*, so that they become in fact chains of cells. In certain members of the higher fungi, parts of the mycelium can become densely aggregated to form resting bodies, *sclerotia*, which may enable a fungus to survive long periods of stress. In other fungi the mycelium can aggregate into strands, often with a hard resistant covering. These strands, known as *rhizomorphs*, frequently enable the fungus to grow from place to place.

The vegetative part of a fungus is usually embedded in the tissues of its host, the only visible parts being those of the fruiting body. With this type of growth the mycelium either enters the plant cells or grows between them producing special structures called *haustoria* which penetrate the cells enabling the fungus to extract nourishment from them.

In general, fungi reproduce themselves by minute unicellular or multicellular spores. Sexual types are formed after the joining together of two cells, and asexual types as a result of mitosis in a cell nucleus.

The classification of fungi is based to a large extent on the nature of the fruiting body in or on which the spores are produced, and on the form of the spores themselves. The commonest classification is into *Phycomycetes*, characterised by having aseptate hyphae; *Ascomycetes*, characterised by having septate hyphae and sexual spores which are borne in 'asci'; *Basidiomycetes*, characterised by having septate hyphae and sexual spores borne on 'basidia', and *Fungi imperfecti*, characterised by having septate hyphae but no distinct type of sexual spores.

BACTERIA. Bacteria form a major group of both disease-causing and beneficial organisms, but as a group they are not so important in agriculture as the fungi. They are very small in size and range widely in shape. They exhibit a large surface area to the air which allows them to carry out rapid chemical changes, and one of their main features is the ability to survive long periods of drought. This makes it possible for them to be transmitted on seeds and bulbs as well as disseminated by the wind, water, slugs and insects. Some of them having tail-like structures called flagella are able to move by themselves.

When attacking plants they usually live in young, soft, sappy tissues rich in amino-acids. They enter plants via natural openings such as lenticels or through wounded tissues. Symptoms may not show up immediately, and in some cases weeks may pass before they become discernible. Once established the bacteria usually live in the plant until its death. The damage to plant material is caused either by enzymes which stimulate the rotting of tissues, or toxic chemicals that kill growth.

Their classification is somewhat complex. Some systems incorporate physiological behaviour such as the ability to ferment different carbohydrates, whilst others are based on the types of cells that they attack. Common classification is into *Parenchyma diseases*, where the bacteria attack parenchyma tissues and cause soft rots; *Vascular diseases*, where vascular tissues are attacked causing wilting of plants; *Systemic diseases*, where vascular and parenchyma tissues are attacked; and *Hyperplastic diseases*, where meristematic tissues are attacked and the production of plant galls is encouraged.

ACTINOMYCETES. Actinomycetes are organisms similar to bacteria which consist mostly of narrow branched threads or filaments. The diseases are immotile and reproduce themselves by breaking off the ends of the filaments to produce spore type structures. The commonest example of this type of disease is the Common Scab of potatoes *Streptomyces scabies*.

VIRUSES. Viruses consist of small particles of nuclear protein. When established in plant tissues viruses are capable of causing disease symptoms, and although they rarely kill their hosts they cause a reduction in growth, distortion or discoloration.

In general, viruses can live and reproduce only in living plant material but examples such as the Tobacco Mosaic Virus are found which can survive for short periods of time in organic waste. Entry into plants is either via wounded tissues or by the touching of clean and infected tissues.

They are disseminated in many ways, some of the commonest being by contact, for example on knives, clothes, machinery and humans; by insects, mites and eelworms; by seeds, although in agriculture this is rare, and by soil transmission, via organic waste or eelworms.

The classification of virus diseases is complex but the system commonly followed is to name the disease after the host it was first found attacking, for example Arabis Mosaic Virus was first found affecting Arabis.

MYCOPLASMAS. Mycoplasmas are living organisms, smaller than bacteria and larger than viruses and they contain nuclear-proteins. They are usually spread by species of Leafhopper and their growth is limited to the phloem tissues of plants. The main type of disease that is produced is that of 'Yellows'.

Plant Protection

Plant protection is a complex study consisting of many inter-relating factors, and although chemicals play a major role in the control of plant disorders, there are many other methods available. The most important means of control are discussed below.

Natural Controls

Climate, physical barriers, soils and food quality are the major natural controls.

Long periods of rain or drought can inhibit the spread of pests and diseases. Temperature is probably of more importance as it affects the speed of development and spread, as well as discouraging the ingress of many harmful species.

Physical barriers such as mountain ranges, deserts, and large water masses all have an influence on the movement of plant disorders. In many cases harmful species are unable to cross these barriers, and their movements are thus restricted. If they do cross, it is important that their natural predators are introduced to the new locations to prevent the outbreak of epidemics.

The acidity or alkalinity of a soil can inhibit the development of pests and diseases. Species of fungi such as Clubroot of brassicas are restricted in acid soils whereas many insects develop successfully only on neutral to alkaline sites. The moisture content of a soil is significant as many pests such as slugs, snails and wireworms are abundant on heavy moist clays, whereas others such as cutworms prefer loose dry conditions. The nutrient status of a soil bears on its susceptibility to disease: excessive soil nitrogen decreases resistance to some diseases whereas adequate amounts of potassium can sometimes enhance resistance. *Trichoderma viride* can be found in some acid soils. This fungus has an antibiotic effect on other pathogenic fungi such as *Fomes annosus*.

Healthy plants are capable of withstanding minor checks to growth or small infections and infestations although it would be a gross mistake to think that healthy growth is synonymous with resistance.

owing are as follows.

rganic waste left lying around farm- and diseases so general cleanliness

ERRATUM

For p. 129 1.25 please read:
... are restricted **to** acid soils ...

is to be encouraged. Plant hygiene, especially when new material is brought onto a holding is of equal importance. Numerous troubles are spread by the unsuspecting grower planting and sowing infected stock. The rule should always be to buy only clean, healthy, certified materials.

CROP ROTATION. Although rotation is not really so important as in years gone by it can still inhibit the build up of harmful organisms, and where skilful rotational practices are coupled with the correct use of chemicals good control can be achieved.

CULTIVATIONS. The correct and timely use of cultivations can lead to the reduction of harmful organisms. Probably the most important example is where weed species which can harbour harmful pests and diseases are ploughed into the soil.

TIME OF SOWING. By sowing crops at specific times a reduction of harmful organisms can be achieved. The work has mainly been carried out on pest species, and typical examples are the late sowing of carrot crops to miss the major hatching period of the Carrot Root Fly, and the early sowing of pea crops which are less susceptible to damage by the Pea Moth.

DRAINAGE. Free-draining soils increase the activities of beneficial soil-borne organisms which rapidly break down organic matter, thus reducing sites of infection. Drainage has the added advantage of encouraging healthy growth which is then less susceptible to attack.

Genetical Resistance

Plant species can now be bred with inherent pest and disease resistance characteristics. Most work has been carried out against diseases but pest control has been achieved in some notable instances. It sometimes happens that a variety of plant highly resistant to a pest or disease is commercially unsuitable because of its low yield potential or unacceptable appearance, although these plants contain a readily usable 'gene pool', where resistance characteristics can be transferred to more desirable plants. Because there is always a chance that the harmful organism will mutate to produce a form unaffected by the resistant genes, resistant varieties should not be grown in close proximity to susceptible types.

Closely related to resistance is 'cleaning up', or removing established diseases from plant materials. This is done by one of three methods: heat treatment; growing in isolation coupled with roguing, or meristem culture. The latter is the production of plants from single or small numbers of healthy cells. These techniques are used against a wide range of virus diseases in commercial and ornamental species.

Legislation

In many countries special precautionary control measures are enforced by law to prevent the introduction of harmful organisms and to eradicate any recently introduced species. They encompass the compulsory inspection and quarantine of plants and animals entering or leaving the country. Diseased plant orders prohibit the sale of material substantially infected with harmful organisms and also limit the areas or soils where susceptible plants can be grown: in certain areas it is forbidden to grow sugar beet in soils infested with Sugar Beet Eelworm.

Biological Control

Parasites, diseases and natural predators can kill or reduce the population of a pest or disease species. This type of control has been used mainly in the control of pest species although as mentioned above the fungus *Trichoderma viride* can inhibit the development of certain other pathogenic organisms. Control is in the form of predaceous or parasitic methods.

Many insects and mites feed by preying on other insects and mites, devouring them, or sucking out their body fluids. Usually each predator can kill large numbers of its host species. Biological control is effected by predators that feed on particular species, or on pests in general.

Parasites feed in or on bodies of their hosts, their eggs commonly being laid inside the host's body. As a general rule each parasite kills only one pest during its life cycle.

Many insects and mites carry out a beneficial reduction of pest species, although other organisms such as bacteria attack the larvae of butterflies and moths, and it is important to remember these agencies when formulating control systems.

Biochemical Methods

Pheromones or sex attractants draw pest species into specially designed traps. Their use at the moment is restricted to calculating the numbers of pest species on a specific site and the timing of sprays can thus be calculated. Another biochemical method of limited use is that of the introduction of sterile males. Large numbers of irradiated male insects are released to mate with females, the females subsequently laying sterile eggs. The use of pheromones and the introduction of sterile males are methods in their infancy, and even when developed further it is unlikely that they will take the place of chemicals.

Chemical Control

The use of chemicals, however great the range to choose from, is no substitute for good husbandry and if chemicals are used it is important for good results that they are applied correctly; that the right material is used at the precise time in suitable weather conditions and with well maintained equipment. They should be able to withstand adverse weather conditions after being applied and be economical to use.

The commonest chemicals are *Insecticides*, used primarily to control insects although many have a dual action; *Acaricides*, formulated specifically to control mites, but, as with insecticides, some control of other organisms can be obtained: *Molluscicides*, designed for the control of slugs; *Fungicides*, used in the control of various species of fungi although some fungicides can have a limited effect on certain pest species, and *Bacteriacides*, developed to control bacteria.

Chemicals work as stomach poisons; by contact; as fumigants or as systemic chemicals which are substances absorbed into plant tissues and thence into the pest or disease.

To constitute effective control chemicals must be applied to ensure maximum and even contact with the pests or diseases. To achieve this, the active ingredient is commonly mixed with other substances, often called 'carriers', to aid its dispersal. The main carriers of agricultural chemicals are water, wettable powders and miscible oils.

The degree of evenness obtained will depend upon such factors as the quality and suitability of the equipment, the skill of the operator and the effect of uncontrollable factors such as wind and rain. To ensure accurate applications frequent static and field calibrations should be carried out. To prevent accidents machinery should be well cleaned after use, and staff should be provided with adequate protective clothing as well as instructed in the procedure to follow if an accident does occur.

Integrated Control

Integrated control is an amalgam of chemical, biological, biochemical and natural systems brought together in such as way as to gain the maximum advantage from each. The system brings together the 'cleaning up' of stock, breeding for resistance, selection of predators and parasites, correct cultural techniques and chemical control. It controls pests more than diseases.

To achieve integrated control on a field scale can be difficult and the main areas of use are in intensive fruit, vegetable and glasshouse crop production. Benefits include reduced labour and chemical costs and a reduction of spray residues and phytotoxic effects.

For the system to be successful it is essential that some pests are present on the crop before any predators are introduced, and in some cases it is necessary actually to introduce pest species at the same time as the predators. This should be carried out at an early stage and an even distribution is essential. A careful watch is necessary to check for isolated outbreaks of the pest.

Chemicals should be used only to control pests and diseases unchecked by predators or other agencies. When they are used systemic chemicals or those which have a low toxicity to beneficial organisms should be selected.

New materials for pest control are always being developed and whilst pests show an ability to build up resistance this will continue. Integrated control is aimed at removing the dependence on chemicals. Its use will become more important as the cost of chemicals increases and use of them is restricted.

PLATE DESCRIPTIONS AND CONTROL NOTES

1 Periods of heavy rain, hail and winds (especially dust laden winds) can result in poor seedling development, reduced growth, serious physical damage and chlorotic tissues (lacking in green colour). As these are natural and largely unpredictable phenomena generalised control measures cannot be adopted. However, it is possible in some circumstances to insure against hailstorms. Also, in soil blowing areas the technique of planting straw between crop rows can be beneficial.

Nitrogen is an essential plant nutrient used in the production of chlorophyll. It is required in large quantities and can be easily leached from soils. Deficiency symptoms appear as stunted and spindly growth, pale green to yellow tissues and a marked reduction in yield. Control consists of balanced applications of nitrogenous fertilisers. The damage can be assessed by soil analysis coupled with knowledge of the past cropping history.

2 Barley grows well on a range of soils but its root system can be susceptible to acidic conditions. In fact, crops can be total failures where the soil falls below a pH reading of 5.8. Where crops are grown in acidic conditions they have a tendency to become chlorotic, stunted, and can in severe cases collapse and die.

Phosphorus is an essential plant nutrient used in root development and energy transfer. It is not leached to any degree, but in situations of high or low pH it can become unavailable to plants. It can also become deficient if soils are continually cropped without adequate replenishment of nutrients. Symptoms appear as reduced seedling growth, bluish-purple stems and leaves, poor root development, yellow streaking of foliage, early ripening and reduced yields. Control measures are similar to those in cases of nitrogen deficiency, namely, balanced applications of phosphates calculated after studying past cropping histories and soil analysis.

3 Potassium is a major plant nutrient required in large quantities by plants. Its main functions are to increase sturdiness, give winter hardiness, and in certain circumstances disease resistance. Most soils, especially those which are light textured, require regular applications of potassium. Cereals have a moderate requirement for potassium, but applications should always be balanced with soil nitrogen and magnesium. This is because potassium can prevent the production of lush growth where excess nitrogen is used, and in circumstances where excess potassium is present magnesium can become unavailable.

Deficiency symptoms appear as yellow of margins in older leaves (commonly called marginal leaf scorch), loss of winter hardiness, brown scorching, loss of yield, and in barley the characteristic 'white tip' to leaves.

Control of Potassium Deficiency should be based on results of soil analysis, remembering that applications must be balanced with both nitrogen and magnesium.

4 Calcium cyanamide is a relatively quick acting nitrogenous fertiliser which can have a caustic effect on plant roots, especially on seedlings growing in sandy soils. Symptoms show up as scorching of foliage, poor development, and in severe cases death. Damage can be prevented by applying the chemical 7–14 days before sowing, ensuring correct application rates, and avoiding contact of the root with the concentrated material.

Magnesium is an essential plant nutrient used in the production of chlorophyll. However, plants require only about one tenth the amount of nitrogen. Magnesium has the ability to move about in plants, and most of it is transmitted to areas of high need, such as young shoot tips. Cereals are not seriously affected by shortage of magnesium but deficiency symptoms can appear as chlorosis starting in the older leaves; yellowing of leaves with darker venous areas, withering of tips in barley; and in severe cases the withering of whole plants.

Magnesium Deficiency can occur where excessive amounts of potassium are present. It is important therefore to balance applications of magnesium with those of potassium. However, symptoms can also appear on plants which are growing in poorly structured soils that are too wet or dry. Symptoms can be alleviated by applications of magnesium limestone, kieserite, Epsom salts or calcined magnesite.

5 Manganese is an essential plant nutrient, but compared to nitrogen, phosphorus and potassium it is required only in small amounts. Deficiencies occur on light land containing excessive amounts of lime. Symptoms vary with the crop: barley suffers from grey speckling and wheat grows limp and yellow with interveinal chlorosis of older leaves.

Control is by applying amounts of manganese sulphate calculated from results of leaf analysis. Soil analysis is ineffective as a means of identification.

6 Symptoms of drought damage can occur in cereals growing on free draining sandy soils: sands quickly release their available water, and during periods of hot dry weather severe moisture shortage can result in dying back of leaf tips, poor growth and reduced yields. Control of the problem is difficult; it consists of building up the soil moisture reserves by applications of organic matter, and the prevention of moisture loss by the appropriate cultural techniques.

Copper Deficiency is uncommon in Britain and is usually restricted to sandy heathland, deep peats and shallow chalks. The Ichnield series of soils are the most badly affected. Symptoms vary with individual crops: in wheat young leaves become white and shrivelled, and in barley young leaves wither and show grey tips. Control measures consist of foliar sprays of copper oxychloride in May or, for longer lasting results, soil applications of copper sulphate.

Foliage applications of dinitro-orthocresol can lead to scorching. Care should always be taken whenever applications are carried out, because any damage is usually permanent.

7 Frost Damage can be caused by wind or by radiation frosts. Wind frosts usually occur during the winter months and in Britain are brought by freezing winds from Siberia. Radiation frosts occur from September to June on clear, still nights. The actual damage is caused by the production of ice crystals in the intercellular spaces. The crystals have a tendency to rupture the cell walls, and when they melt the resultant moisture cannot return to the cells quickly enough. This means that the plants can in effect be drowned in their own sap. Control of Frost Damage is by producing strong sturdy growth by balanced applications of fertilisers, and by the selection of the correct varieties to suit specific conditions.

8 Boron is an essential plant nutrient required in small quantities. Deficiencies occur in light soils containing large amounts of lime and the trouble is exaggerated during periods of drought. Control can be achieved by applications of borax, the rate being decided after soil

analysis. Excess boron compounds are extremely poisonous to plants and great care should be taken in their application.

Iron is another essential nutrient required in small quantities. Most soils contain enough iron, but deficiency symptoms can occur where excess calcium, magnesium, zinc or chromium are present. Symptoms appear as chlorisis of tissues, the areas of damage being either completely bleached, or having a mottled appearance. There is no really economic control of Iron Deficiency but in certain circumstances the use of chelates may be considered. The usual alternative is to avoid growing susceptible crops on soils with a history of Iron Deficiency.

Sodium chlorate is a total weed killer used to control a wide range of weed species. However, over application or seeping of the chemical can cause severe damage to crops. Unfortunately this material can remain active for as long as twelve months and therefore the recovery of damaged plants is impossible.

Sea Salt Injury to plants is either in the form of physical damage caused by rapid moisture loss and abrasion by salt particles, or by scorching of foliage caused by excess salt concentrations on the surface of leaves.

9 Halo Blight is a bacterial disease which is transmitted on seeds and by the splashing of rain droplets. Symptoms show first as small spots on the leaves of seedlings, but these later spread to the older leaves, glumes and leaf sheaths. The severest damage is seen on the younger leaves. Care should always be taken in identification, as symptoms can be similar to those of Manganese Deficiency and Leaf Spot. In severe cases the central tissues of leaves become pale green and shrivelled, but attacked tissues always show distinct yellowish halos. Specially treated seeds can be bought which should be immune to this infection.

Leaf Spot is caused by the fungus *Septoria avenae*. Symptoms appear as round or oval yellowish-brown spots which contain numerous black pycnidia (structures which allow the fungus to survive adverse conditions). Control consists of using benomyl-thiram seed treatments.

10 Take All is caused by the fungus *Gaeumannomyces graminis* (*Ophiobolus graminis*) and in certain circumstances it can be a serious disease of wheat. The fungus survives on the stubble of previous crops, and on grass roots, which can make it a problem where continuous cropping is practised. It can also cause serious damage on light alkaline soils or sites that are badly drained, compacted, or over limed. However, it is uncommon on heavy fertile soils. Symptoms appear in two forms, namely, the 'Take All' stage where patches of young plants are severely stunted or killed, and the 'Whitehead' stage which results in empty bleached ears. Adequate rotation of crops, maintenance of high soil fertility, reducing perennial grass weeds and a high level of husbandry should control this fungus.

Black Mould is not an active parasite of cereals and in fact its appearance is usually an indication that the crop has been affected by some other trouble such as Take All or Potassium Deficiency.

11 Eyespot is caused by the fungus *Pseudocercosporella herpotrichoides* (*Cercosporella herpotrichoides*). It is a serious disease of winter wheat and barley, with rye as an occasional host. The fungus favours mild weather in winter and spring, and severe attacks can occur in thick stands of soft, lush growth. Symptoms appear as oval, brown bordered smudges on young leaf sheaths which later develop into eye shaped areas often with a black centre. Control involves a two year grass break or three year arable break from wheat

or barley, coupled with the use of resistant varieties and chemicals such as thiophanate-methyl and carbendazim.

Sharp Eyespot (*Rhizoctonia cerealis/Corticium solani*) is similar in appearance to Eyespot, the main difference being that the lesions are more defined and angular, whilst the dark borders are easily distinguished from the linear areas. It is not a very serious problem in Britain but at present no chemicals are available for its control.

12 Foot Rot, Ear Blight and Oat Cap are caused by various soil borne species of *Fusarium*. Damage occurs mainly on cereal roots growing in heavy, poorly drained acid soils. The fungi attack young seedlings which can be killed outright, resulting in thin stands. On older plants symptoms appear as yellowish leaves with withered tips, poorly developed root systems, whitish pink spores at the base of stems and the failure to form ears. In the later stages spores can spread to the ears to produce Ear Blight. The symptoms of Ear Blight are bleached spikelets, the formation of white mycelia and masses of pink spores. The purchase of treated seed and the use of benomyl and thiophanate-methyl will give some control.

Snow Mould caused by the fungus *Fusarium nivale* occasionally attacks crops of wheat and rye in Britain. Damage appears after snow on soils containing excessive nitrogen, and can cause the rot of large groups of plants.

13 Leaf Spot caused by the fungus *Septoria tritici* occasionally attacks crops of wheat and rye producing light coloured spots covered in dark spore cases. It causes only minor trouble in Britain.

Various races of Cereal Mildew (*Erysiphe graminis*) can attack wheat, barley, oats and rye. When serious infections occur, yields can be reduced by up to 20 per cent. Damage tends to be worst on susceptible varieties sown in late spring on soils of very high fertility. Symptoms appear as white patches on the surface of leaves, sheaves and ears. Attacks frequently appear first on the lower leaves as dirty white masses of mycelia and as the crop develops the fungus spreads upwards. Control of the disease consists of growing resistant varieties, and the use of chemicals such as ethirimol, tridemorph and triadimefon.

Ergot caused by the fungus *Claviceps purpurea* is commonly found infecting cereals and grasses. The damage to plants is not serious but the 'ergots' are poisonous to man and animals. Various races exist, attacking mainly rye, but wheat, barley, oats and wild grasses can be affected. The disease attacks only the flowers. It spreads by spores and once established honeydew, a sticky substance, exudes from the florets. This contains spores which further spread the infection. The ergot itself is a horn shaped purplish-black structure which develops in the ears at the expense of some grain. Ploughing in overwintering ergots, crop rotation, and buying clean seed will provide control.

14 A number of soil borne members of the genus *Pyrenophora* cause damage to cereals. They tend to be found in wet, cold conditions, where they all produce similar symptoms. All the species are seed borne. The best means of control is to purchase specially treated seed.

Leaf Blotch of barley caused by the fungus *Rhynchosporium secalis* produces scald-like lesions on the leaves, leaf sheaths and ears. Infection is erratic but occurs via infected seeds, old stubble and certain weed grasses. It is controlled by growing resistant varieties, buying treated seed and applying chemical sprays such as benomyl, captafol and thiophanate-methyl.

15 The Rusts are a very important

group of disease-causing organisms belonging to the basidiomycetes. The commonest species which attack cereals are Crown Rust (*Puccinia coronata*), Brown Rust (*P. hordei*), Yellow Rust (*P. striiformis*) and Black Rust (*P. graminis*). They are commonly found on leaves and stems, but attacks usually cannot be seen until the spore stages are produced. Symptoms appear as various masses of coloured spores which can cripple plants, but rarely kill them. Rust life cycles are complicated, with up to five types of spore being produced at different times of the year, and in some cases on different type of plant. However, not all exhibit each spore type. The 'autoecious rusts' spend all their life cycle on one host, whilst the 'heteroecious rusts' spend part of their life cycle on totally different species. Control measures involve growing resistant varieties, stubble cleaning and the use of chemicals such as benodanil, triadimefon and oxycarboxin.

16 As mentioned in Section 15, heteroecious rust species spend part of their life cycles on more than one plant. Two examples of this type of life cycle are Crown Rust *Puccinia coronata*, where part of its cycle is spent on oats the other part on common buckthorn *Rhamnus cathartica*, and Black Rust *P. graminis* which spends part of its life cycle on species such as couch grass *Agropyron repens*, the other part on common barberry *Berberis vulgaris*.

17 A number of parasitic smuts which affect cereals can be recognised by the black soot-like spores usually produced in the ears and grain of plants.

Stripe Smut of rye, caused by *Urocystis occulta* can result in stunting, malformation and streaking of leaves and stems. It is not of serious economic importance, but control can be achieved by seed treatment and crop rotations.

Bunt of wheat caused by *Tilletia caries* has declined in seriousness. It can be recognised only at ear emergence where infected ears are longer and narrower than healthy ones and have black masses of greasy spores. The use of treated seed provides good control.

Loose Smut caused by *Ustilago tritici* can cause damage to wheat crops. Its life cycle is similar to *Ustilago nuda*. See Plate 18. It is controlled by specialised seed treatments.

18 The fungi known as Smuts are a large group belonging to the Basidiomycetes. Their name is derived from the characteristic production of black sooty spores. They are all obligate fungi, living on a wide range of species, and in certain cases they can cause serious loss of yields. Control of Smuts can be rather difficult as many species produce differing biological races. However, control can be obtained by seed treatments, and certain varieties show resistant characteristics.

19 The Saddle Gall Midge *Haplodiplosis equestris* can cause economic losses on cereal crops. Infestations are found on most cereals and other grasses, damage being worst on winter wheat and barley grown on heavy land. However, high soil temperatures in early summer coupled with moist conditions can lead to high populations.

Symptoms are caused by the larvae feeding just above the nodes near the tip of the plant. Damage appears in the form of sunken saddle shaped galls which can join together during high infestations. Control measures include substituting non-cereal crops into barley and wheat rotations, early sowing, the control of weed grasses, and the use of chemicals such as fenitrothion.

20 Cereal Cyst Eelworm *Heterodera avenae* is a troublesome pest, especially of oats. It is common in the southern

counties, north and east Midlands of England. Symptoms of attack appear as stunted patches of yellowish plants in otherwise healthy crops; the effects look similar to Nitrogen Deficiency. On closer examination the plants are found to have shallow matted root systems composed of stunted and dead rootlets. The cysts which are soil borne are dark brown lemon shaped structures containing many young Eelworms. However, when attached to plants they have a typically whitish tinge. The cysts have the ability to survive in soil for several years. To control this pest avoid excess cropping of oats, keep soil fertility high, control wild oats, and grow resistant varieties.

Stem and Bulb Eelworm (*Ditylenchus dipsaci*) is widely distributed and attacks a range of plants. Eggs are laid inside the plants and when the young hatch, they feed causing the production of distorted shoots and yellow speckles on the foliage. Control consists of using resistant varieties, crop rotations, stubble burning and weed control.

21 Many species of aphid attack cereals, and when large infestations occur serious loss of yield can result. The aphid's life cycle is spent between two hosts. In spring to early summer they leave their winter hosts and migrate to soft, rapidly growing plants. They feed by piercing the tissues and extracting the cell sap. The symptoms of damage are varied and can appear as stunting of growth, poor colour, spotting of leaves, blindness, shrunken kernels, sooty moulds and, although not a direct symptom, the spread of virus diseases. Control consists of applying chemicals such as chlorpyrifos, dimethoate, phosalone and pirimicarb.

Thrips or Thunderflies are small insects which attack a range of plants. They feed by piercing and scraping tissues, the damage appearing as malformed shoots, stunting and mottled bleached areas. Control can be achieved by the same methods that apply for aphids.

22 Wireworms are very common soil inhabiting insects which are capable of causing severe damage to cereals. They are the larvae of the insects known as Click Beetles, the commonest species of which are members of the genus *Agroites*. Damage is caused by the larvae hollowing out grains, severing young stems from their roots, and tunnelling into the base of shoots of established plants. Symptoms appear as death of cereals in patches, yellowing foliage, and general slow development. Control can be achieved by applying benzene hexachloride (HCH) to the soil.

Flea Beetles attack a wide range of plants, the main damage being caused by the adults eating holes in the leaves and stems of seedlings just above ground level. The damage is most serious where large infestations are present in dry springs. Control consists of the maintenance of hygiene and use of chemicals as for Wireworms. For Chafer control see Plate 69.

23 The Wheat Stem Sawfly *Cephus pygmaeus* is a destructive pest of wheat and other grains. The larvae feed inside the plant stem starting from the shoot tips and working towards the stem base. Once near soil level they completely eat out the centre of the stem causing the plant to break off. Control measures involve burning or ploughing in old stubble, growing non susceptible crops such as legumes, and the use of resistant varieties.

The larvae of the Common Rustic Moth *Mesapamea secalis* is mainly a pest of grasses like *Dactylis spp*, *Poa spp* and *Festuca spp*, although damage can be seen on wheat and oats occasionally. The larvae bore into the bases of stems causing similar damage to that of Frit Fly. No control measures are usually warranted.

24 Bibionid Flies *Bibio spp* are primitive insects which sometimes cause minor damage to cereals. They feed on roots of plants, the damage becoming visible during February and March.

Gall Midges *Contarinia spp* are small delicate flies which frequently attack plants. Symptoms can be very varied, ranging from leaf distortions to the production of galls. The Wheat Midge *Contarinia tritici* attacks the flowers of wheat causing the flattening of spikelets and a reduction in grain size. Since attacks are sporadic, no control measures are usually practised, although where necessary numbers can be reduced by using DDT.

Leatherjackets are the larvae of insects commonly called Crane Flies *Tipula spp*. They are important soil borne pests of many agricultural crops, and are most common in damp summers and autumns although the most serious attacks occur in spring. Control can be achieved by using chemicals such as benzene hexachloride (HCH).

25 Frit Fly larvae can cause considerable damage to cereals, especially where crops are sown after grass, or where stubble has been left. The larvae bore into stems of cereals which causes the central shoot to turn yellow, wither, and eventually die. The main control is obtained by using chemicals such as benzene hexachloride (HCH) and chlorfenvinphos.

26 The Wheat Bulb Fly *Leptohylemyia coarctata* is prevalent in wheat growing areas of Britain. Attacks are usually restricted to winter wheat, although other cereals like winter barley can also be damaged. The larvae attack the central shoots of plants causing the leaves to turn yellow, and the eventual death of the growing points. Control is obtained by using seed treated with chemicals such as chlorfenvinphos, or by spraying crops with dimethoate.

Larvae of the Cereal Leaf Beetle *Lema melanopus* damage crops by eating longitudinal strips from the leaves. Control measures are unnecessary.

Evidence of Leaf Miner *Hydrellia griseola* appears as white longitudinal areas on cereal leaves. Again, no control measures are usually necessary.

27 The Oat Spiral Mite *Steneotarsonemus spirifex* feeds in large numbers on oat crops. Damage appears as a typical reddening of tissues, the formation of gall-like structures beneath the sheath leaves, stunting of growth and twisting of the rachis. Control measure are not warranted, although crop rotations inhibit the development of large populations.

28 Mice can sometimes cause damage to cereal plants, however damage is rarely of any significance and control is uneconomic.

Birds such as sparrows and geese can cause severe damage to certain crops. Damage appears as pecked holes and occasionally the severance of plants close to ground level. Control using bird scarers may help, but in many cases control is uneconomic.

Slugs can cause serious damage to cereals and other crops. The main crops to suffer are autumn sown cereals and main crop potatoes. Slugs are nocturnal creatures feeding in moist soil throughout the year. Symptoms appear as bare patches in fields, eaten shoots, leaf shredding and holes in the base of stems. Control can be achieved by producing fine seed beds at sowing time and the use of chemicals such as methiocarb and metaldehyde in made-up baits.

29 Frost Injury can occur on most crops during the winter months. Damage can

be serious especially where soft lush growth has been produced by excess application of nitrogenous fertilisers. See also Plate 1.

Mildew *Erysiphe graminis* is a serious disease of cereals and grasses. For notes on symptoms and control see Plate 13.

Feeding Ergot contaminated cereals and grasses to animals can cause them severe injury. The disease attacks the flowers of the Gramineae, the degree of infection being dependant upon climatic conditions at flowering. The amount of Ergot in grasses can be reduced to some degree by intensive grazing and occasional topping; for further notes see Plate 13.

Leaf Fleck *Mastigosporium spp* attacks Timothy grass, causing a spotting of foliage. The spots appear as light coloured flecks with a distinct dark margin. Another disease of grasses is caused by the fungus *Epichloe typhina*. Attacks are mainly confined to Cocksfoot, the symptoms appearing as blindness of inflorescences and light speckling of the foliage.

30 Yellow Slime *Corynebacterium rathayi* of Cocksfoot is an occasional disease in Britain. Symptoms are seen where infected seed is sown. The damage appears as bare patches in crops, and plants covered with yellow slime. The use of healthy seed should prevent attacks.

The Smut Fungus *Ustilago spp* attacks tall grasses transforming the florets into masses of spores.

Timothy Fly larvae *Amaurosoma spp* feed on developing flower heads before they emerge from the leaf sheaths. Damage is usually slight and control can be achieved by timely application of DDT or dimethoate.

Cereal Root Eelworm *heterodera avenae* attacks wheat, barley and some grasses such as ryegrass. Symptoms appear as the formation of cysts on the roots of plants, followed by the stunting and yellowing of tissues. See also Plate 20.

31 Grasses can be severely damaged by various species of Rust. Examples are Crown Rust *Puccinia coronata* on ryegrass, Meadow Grass Rust *P. poarum* on *Poa spp*. Brown Rust *P. recondita* on brome grass and Yellow Rust *P. striiformis* on Cocksfoot. The symptoms appear as the production of characteristic rust spores at different times of the year. Control can be achieved by chemicals such as maneb, nickel sulphate or systemic fungicides.

Gall Midges *Mayetiola* and *Oligotrophus spp* can cause damage to various species of grasses. Symptoms range from the production of swollen stems to the reduction of grain development. Control measures are usually unnecessary.

32 A wide number of pests can cause damage to grasses in Britain. Some examples are the Meadow Plant Bug *Leptopterna dolobrata*, the Flounced Rustic Moth *Luperina testacea*, the Timothy Tortrix Moth *Amelia paleana*, the Hessian Fly *Mayetiola destructor*, and the Tarsonemid Mite *Siteroptes graminum*. The symptoms of attack vary with the individuals, ranging from the severing of stem bases to the destruction of the developing seed.

Control measures are often unwarranted, but when necessary, outbreaks can be reduced by timely applications of insecticides or acaricides.

33 Chlorosis of plant tissues may be caused in a variety of ways such as virus attack or nitrogen and/or magnesium deficiency. It shows as the yellowing of normally green leaf tissues. Remedial measures consist of identifying and controlling the individual causal problems.

Necrosis is often a symptom of plant diseases. It consists of the partial discoloration and death of leaves, stems, shoots and fruits.

Potassium Deficiency in clover produces whitish spots, marginal yellowing and browning of the leaves. Reduction in yield can result, but it must be remembered that if potassium fertilisers are applied they must be balanced with soil nitrogen and magnesium, or poor growth can result. (See Plate 3).

34 Leaf and stem discoloration of clover can be caused by nutrient deficiencies where for example a shortage of magnesium leads to chlorotic growth and a distinct brown margin to the leaves; by viruses and mycoplasmas which cause mosaic symptoms on leaves or phyllody of inflorescences or by Leaf Spots caused by pathogens such as *Polytrinchium trifolii*, *Stemphyllium sarciniforme* and *Kabatiella caulivora*.

In the case of nutrient deficiencies, control can be achieved by applications of the deficient nutrient. However, the control of the diseases is more complicated, because sometimes it is very difficult to assess the economic damage caused by the individual species.

35 Proliferation, or the continued formation of new growth beyond normal limits, is common in clover. It leads to inflorescences becoming densely compacted with flowers, or leaves arising where they do not usually occur.

Mildew *Erysiphe polygoni* is a variable species (often split into distinct races) capable of attacking a range of plants such as peas, swedes and clover. Symptoms appear as a white covering of mycelia over the leaves of clover plants which can reduce vigour; in certain other species yield is affected.

Clover Rot caused by the fungus *Sclerotinia trifoliorum* is a common disease where clover has been grown for long periods. The most serious damage is caused to Broad Red Clover, but attacks have been found on most species, and also on field beans. Damage appears during a mild winter following a wet autumn. The symptoms are the discoloration of foliage, flagging (wilting), and death of isolated clumps. It is controlled by using clean seed, practising adequate rotations, avoiding clover on infected land for eight to twelve years, and applying balanced nutrients, especially calcium, potassium and phosphorus.

36 Rust of clovers caused by the fungus *Uromyces trifolii* can cause minor damage to crops. Symptoms appear as elongated brown spots on stems and circular brown spots dotted on the leaves. In severe attacks whole plants can become shrivelled and stunted.

The Common Earwig *Forficula auricularia* is a general feeder causing damage to a range of plants. However, it is usually more troublesome in horticultural situations than on large scale cropping. Damage to clover is caused by the Earwigs chewing small holes in the leaflets, although in bad attacks skeletonising of the leaves can take place.

Attacks to crops by the Grey Field Slug *Agriolimax agrestis* can cause serious damage. Most losses take place when young seedlings are attacked, but reduction in vigour can result when older leaves are damaged. Attacks can be reduced by producing firm seedbeds and applying chemicals such as metaldehyde or methiocarb.

37 Stem and Bulb Eelworm *Ditylenchus dipsaci* attacks many crops, but in clover causes the symptoms commonly known as Clover Sickness. Damage by this pest can cause serious losses in Red Clover crops. Symptoms in seedlings occur as swellings just below the cotyledons; in a mature crop thin patches containing stunted plants with swollen stem bases are found. Control can be achieved by adequate crop rotation, efficient control of susceptible weeds, the use of resistant varieties such as Norseman, and the purchase of methyl bromide fumigated seed.

38 Clover inflorescences can be attacked

by various species of Weevil, the commonest of which are *Apion aestivum* and *A. apricans* though other species can also cause significant losses. Damage is caused by the larvae feeding on the developing ovules and seeds of all varieties of clover, but some minor damage to foliage can occur by adults feeding. Control is obtained by spraying headlands during April and May with DDT or malathion.

The Clover Leaf Weevil *Hypera nigrirostris* causes minor damage to clover by feeding on the stems, leaves, buds and seeds. They feed openly on the plants, but are difficult to see because of their colour. Where necessary they can be controlled by sprays of DDT or HCH before the flower buds form.

39 Pea and Bean Weevils *Sitona spp* are common pests of leguminous plants. The adults feed on the leaves, causing characteristic leaf damage symptoms (see Plate 39) and the larvae feed on roots and nodules. Damage to seedlings can be serious especially on poor soils and in cold weather. Conversely, older plants can withstand a severe hatching without any appreciable loss of yield. The main problem with this pest is that it can carry two virus diseases, namely, Broad Bean Stain and Broad Bean True Mosaic; even so, chemical control is not usually justified.

Frost injury to lucerne produces typical whitening of leaf margins. However, as the crop is relatively hardy no serious loss of yield results.

Potassium Deficiency in lucerne appears as whitish spots on the leaves as well as marginal yellowing and browning of tissues. Large quantities of potassium may be required to rectify this deficiency but the rate of application should be calculated in the light of past experience and soil analysis.

40 In severe cases Boron Deficiency can cause the death of apical meristems. Minor symptoms appear as the formation of distinct light green or brown leaf margins and reduced plant vigour. Control is obtained by soil applications of borax.

Anthracnose *Colletrotrichum trifolii* causes the wilting and death of lucerne plants. Affected plants have dry foliage which easily wilts and shrivels. Lesions are produced at the base of plants, and when the stems are completely encircled they die.

Verticillium Wilt Disease *Verticillium albo-atrum* has become a serious problem in lucerne crops. Affected plants show a distinct yellowing of the leaves from the base of the plant upwards, followed by wilting, shrivelling and eventual death. The first symptoms appear in early summer and become more severe as the season progresses. Resistant varieties and seed dressed with thiram will provide control for this disease.

41 Leaves of lucerne affected by Black Stem Fungus *Ascochyta imperfecta* appear slightly yellow and have irregularly shaped spots with distinct dark margins. The spots may join to produce larger infected areas. Stems can also be infected causing cankering and the eventual death of shoots. The disease is seed borne and can be controlled by using thiram seed dressings.

Downy Mildew *Peronospora trifoliorum* is a minor disease of lucerne causing the formation of violet-grey mycelia over the leaves. Control is unwarranted.

For Clover Rot *Sclerotinia trifoliorum* see Plate 35.

Severe attacks of Lucerne Leaf Spot *Pseudopeziza medicaginis* are uncommon in Britain, but where they arise defoliation of plants can result. Leaf symptoms appear as dark brown to black spots scattered over light-coloured leaves, the worst damage being seen during wet autumns.

Outbreaks of the Gall Midge *Contarinia loti* are rare in Britain, but they can occasionally attack *Lotus spp* causing deformed flowers. No control measures are necessary.

42 The Stem and Bulb Eelworm *Ditylenchus dipsaci* is an important pest in Britain. It has a range of hosts, and a number of distinct races exist. Symptoms are seen as the swelling of stem bases, the formation of galls on seedling stems, development of the basal buds, and distortion or stunting of foliage. Damage can be prevented by growing resistant varieties and having seed treated with methyl bromide.

Lucerne Leaf Weevil *Hypera variabilis* is a minor pest of lucerne in Britain. The adults feed on leaves and stems, whilst the larvae attack stems, flowers and seeds. Damage is usually noticed during flowering, by which time control measures are ineffective. Seriously affected crops should be cut as late as possible to remove the larvae with the crop, and chemicals such as DDT and HCH can be used before the flower buds develop.

43 The roots of lucerne plants are sometimes eaten by the larvae of Root Weevils *Otiorrhynchus ligustici*. The roots are gnawed away and this is followed by yellowing of foliage, wilting and eventual death. Where control is necessary the larvae can be killed by applications of HCH or DDT.

The Lucerne Flower Gall Midge *Contarinia medicaginis* causes damage to the flowers of lucerne and related species. This subsequently leads to the reduction in the amount of seed produced. Control can be effected by applying DDT before the flowers are formed.

In addition to the Flower Gall Midge, lucerne and trefoils can be attacked by a number of species of Leaf Midges such as *Jaapiella* and *Dasyneura spp*.

44 Powdery Mildew *Erysiphe polygoni* is rarely troublesome in Britain except in very hot dry summers and occasionally on late sown crops. No economic control measures have been formulated.

Grey Mould *Botrytis cinerea* can be serious in dull, wet seasons, causing discoloration and the production of a characteristic grey dust-like mould over the surface of the pod. In severe attacks this can spread to the seed causing a reduction in yield. When serious attacks occur, control can be achieved by applications of fungicides such as benomyl. See also Plate 47.

Leaf and Pod Spot caused by the fungus *Ascochyta pisi* can cause serious damage to seedlings, but damage is more commonly seen on adult plants. Symptoms appear as pale brown or smaller bluish brown spots on the leaves, stems and pods. These spots soon become sunken, blackened cankers and in severe cases the seed becomes discoloured. Infected seed is a common source of infection in new crops, and the recommended control involves the use of clean seed samples.

45 Pea Rust *Uromyces pisi* is a heteroecious species, part of its life cycle being spent on peas and the other on *Euphorbia cyparissus*. It is rare in Britain.

Pea Aphids *Acyrthosiphon pisum* can cause severe damage and loss of yield in pea crops. The symptoms appear as poorly developed, distorted pods, small peas, and the production of honeydew which can be attacked by sooty moulds. Visual inspection will show the aphids feeding on the undersurface of leaves and flowers, and in warm weather large populations can develop rapidly. Control consists of using chemical sprays such as demephion, demeton-S-methyl, malathion, phosphamidon and mevinphos.

Pea Thrips *Kakothrips robustus* are small insects commonly known as Thunderflies. They cause a silver mottling on young pods and leaves, and in severe cases some contortion of growth can occur. The main attacks occur from early June onwards, and in serious outbreaks a reduction of quality and yield results. Control is obtained by the use of chemicals such as azinphos-methyl, sulphone and fentitrothion.

46 The Pea and Bean Weevil attacks crops of leguminous plants. The main symptom of attack is the serration caused on leaf margins. See also Plate 39.

Pea Moth *Laspeyresia nigricana* is a serious pest of peas and related crops. Damage is caused by the larvae tunnelling into the pods and feeding on the developing seeds. Only relatively small numbers need be present in a crop to drastically reduce its economic value. Pea varieties range in their susceptibility to attack, but most crops in flower or pod during June and July may be attacked. Control is provided by chemical sprays such as carbaryl or fenitrothion.

Leaf Spots and Scorching caused by species of *Ascochyta* can occasionally reduce growth in pea crops. For a general description and control see Plate 44.

47 Grey Mould *Botrytis cinerea* can attack almost any plant at any time of the year, but it is not capable of directly attacking healthy tissues. When it enters a plant it produces a progressive decay of tissues, the cells being killed by an enzyme produced by the fungus. The symptoms show as rotted tissues covered with a greyish brown mould. For control see Plate 44.

Leaf Blotch caused by *Ceratophorum setosum* produces black spots or blotches on leaves and pods, which later turn brown with a lighter green zone at their margin. Large infections can completely kill leaves, but attacks can usually be controlled by removing infected tissues or spraying with copper based fungicides.

48 Root Rot of lupins is caused by various species of fungus including *Thielaviopsis* and *Sclerotinia*. Attacks usually occur on old, weak plants especially in poor soil conditions.

A number of virus diseases can attack field beans. With mosaic virus, symptoms appear as dark green banding along the veins, with light green or yellow chlorotic interveinal areas. The best means of control are to grow resistant varieties and to provide efficient control of Aphids.

The Black Bean Aphid *Aphis fabae* or 'Blackfly' as it is commonly known, can seriously reduce yields in a range of crops. The symptoms occur on beans during June and July. The damage is caused by the aphids sucking large quantities of sap and injuring tissues during feeding. However, damage can also occur on crops because of virus transmission. Control can be achieved by applications of systemic insecticides such as dimephion and thiometon. However, care must be taken to avoid harming pollinating insects.

The larvae of the Lupin Seed Fly *Hylemya florilega* damage plants by feeding on the leaves and stems. The damage is of minor importance in Britain.

49 Potassium Deficiency in mangels shows on the older leaves first. The leaf margins turn pale prior to being scorched brown. The leaf stalks can also turn yellow and exhibit brown streaking. Nitrogen Deficiency on the other hand, causes poor growth, yellowing of leaves, and eventual reduction in yields. Both deficiencies can be assessed by soil analysis, and alleviated by applications of fertilisers.

Magnesium Deficiency of sugar beet produces interveinal chlorosis in the older leaves, followed by shrivelling of foliage and reduced yields. Affected plants may also be infected with secondary organisms such as *Alternaria spp*. Deficiencies can be corrected by applications of Epsom salts or other fertilisers containing magnesium.

50/51 Boron is an essential nutrient in plants, though it is only required in minute quantities. Most soils contain adequate amounts of this element, but in certain circumstances such as alkaline soils or where heavy applications of lime have been made, deficiencies can occur. Symptoms appear as the death of apical buds, sometimes followed by the stimulation of lateral shoots, which in turn also die. Young leaves become blackened and shrivelled whilst the centre of roots become rotted. In beet, the symptoms can appear as a canker round the middle of the root, and the dark affected tissues can be seen when it is cut open. Control measures involve the application of boron either as specially formulated fertilisers, or as a spray of borax at 22 kg/ha before drilling.

52 Manganese Deficiency in mangels appears as typical yellowing of the leaves and is commonly called 'Speckled Yellows'. The leaf margins tend to turn upwards, giving a rolled appearance, and the whole leaf looks triangular in shape. The problem can be remedied by applications of manganese sulphate.

Chlorate injury of beet shows a variety of symptoms. Leaves may exhibit yellow speckling, similar in appearance to Beet Mosaic, and in some cases a slight rolling of the leaf edges can take place.

Seed samples are often treated with chemicals to withstand attacks by pests and diseases. However, when the chemicals are applied at too high a rate, scorching of the young roots can occur. Similar damage can be caused by heavy applications of nitrogenous fertilisers on germinating seeds.

53 Frost Damage can severely reduce yields in a wide range of crops. In beet, symptoms vary with the age and type of tissues affected. The young leaves develop necrotic areas, especially near the tips, whilst the older foliage shows necrosis coupled with a silvering caused by the epidermis coming away from the other leaf tissues. Affected roots can become cracked, and show internal staining, although after severe damage they may turn brown and be invaded by secondary disease organisms.

54 Lightning Damage in beet crops has on occasion been mistaken for symptoms of docking disorder. Lightning may cause leaf scorch, longitudinal and concentric cracking of leaves, root stunting and fanging. Severe damage will give rise to bare patches in fields.

Frost Injury to young seedlings can produce thickened, curled cotyledons whereas in older plants browning and the eventual rotting of tissues can occur. However, care in identification must be taken because the symptoms can be similar to Aphid Damage and Boron Deficiency.

Hail Damage of beet causes torn leaves, sometimes only the main veins remaining intact, but in other cases symptoms can appear as crushed, bruised or tattered tissues. Since damage from these natural phenomena is spasmodic, it is difficult to specify control measures. However, it is possible to insure against Hail Damage.

55/56/57 Viruses are minute structures which can reduce crop yields considerably. Their symptoms are very varied, and care must be taken in identification because they can be mistaken for damage caused by nutrient deficiencies and insects. Viruses may be spread by animals, insects and mechanical damage; the methods varying with each specific virus.

A number of viruses affect the growth of sugar beet, probably the most important of which are Beet Yellows Virus, Beet Mild Yellowing Virus and Beet Mosaic Virus. Each virus causes a reduction in yield, and each exhibits characteristic symptoms.

The main vector of virus diseases in beet is the Peach Potato Aphid *Myzus persicae*. The aphid feeds on infected plants and picks up the virus on its stylet (sucking mouth parts). The length of time the virus remains active in the stylet varies, some types being capable of surviving long periods, whilst others may survive for only a number of hours. The initial spread of viruses occurs by winged aphids feeding on seed crops or mangel clamps, and later moving to field crops. The dissemination of viruses usually reaches its peak in July, and gradually tails off thereafter.

Virus Yellows is an important disease of sugar beet causing severe reductions in yield each year. It is widely distributed, but is not seed borne. Symptoms of the disease appear approximately 10 to 14 days after infected aphids have fed on the crop. However, under certain circumstances this period can be considerably longer. Symptoms appear as the yellowing of the tips of middle age leaves, and in severe cases, the older outer leaves may also be affected. However, the heart leaves will remain green, or show vein etching. On the older yellow leaves small brown or red spots are often found and the leaves become brittle.

Beet Mild Yellow Virus produces orange or yellow leaves with few if any red or brown spots. However, plants infected with this disease are more frequently attacked by secondary parasites than plants suffering from Virus Yellows.

Beet Mosaic Virus appears as a number of small light green flecks on the youngest leaves, and less frequently as a clearing of the veins. The light green areas sometimes appear to have a darker background, but in no circumstances do the leaf tissues show necrotic spots.

Virus diseases in beet can be controlled in a number of ways including chemical control, where insecticides are applied to control aphids on growing crops. Timely applications are essential and the chemicals used must have either a contact or systemic action. It is also possible to achieve control in late sown crops by treating the seed with menazon. Available chemicals include demeton-S-methyl, formothion and phosphamidon.

Good management involving well timed cultural techniques such as late sowings and the production of high plant populations can reduce the incidence of disease.

Crop hygiene ensuring general cleanliness will reduce the initial sources of infections and will prevent the spread of diseases.

In addition to viruses beet leaves can show discoloration because of Nutrient Deficiencies, Pest Damage and Albinism (white areas due to lack of chlorophyll). Where any doubt in diagnosis occurs throughout Britain the local fieldsman of the Agricultural Development Advisory Service should be able to help.

58 Crown Gall *Agrobacterium tumefaciens* is a soil borne bacterium which produces galls and swellings above and below ground level on a wide range of plants. However, many other agencies can cause similar disorders and specialised techniques may be needed for definite identification. The galls vary in size, being whitish, rounded swellings when young (sometimes mistaken for callus tissue), and becoming harder and darker with age. The bacterium enters through wounded tissues and once established it stimulates the plant tissues to grow and produce the characteristic galls. This is not really a serious disease of field crops, but where necessary, attacks can be reduced by miminising

physical damage and burning infected stocks.

Plate 58 B–D show Tail Rot of sugar beet, caused by a bacterial disease and E–F show a Root Rot disease of beet caused by a combination of poor growth conditions and fungal attack.

59 Common Scab *Streptomyces scabies* is a soil borne organism which affects potatoes mainly, but radishes, beet and mangolds are also susceptible. Attacks are mostly seen in dry seasons, on light sandy or gravelly soils where the pH is neutral to alkaline. On beet the symptoms appear as scab-like lesions on the developing roots. Other species of *Streptomyces* and *Actinomycetes* can produce growths and scabs of similar nature. See Plate 59. It is also important that damage from these organisms is not mistaken for physical damage or natural cavities produced during normal growth.

60 Root Rots (Black Leg) of beet seedlings can be caused by a number of fungi such as *Phoma betae*, *Corticium solani*, and *Pythium spp*. They are either true parasites or facultative parasites (being able to live on dead tissues for periods during their life cycle.) These Root Rots give rise to dark constricted areas on young stems either at or near ground level. Heavy infection can occur when the soil is very moist, or if the sowing rate is too high. Control measures involve good soil management and the use of treated seed.

Plate 60 H–L illustrate symptoms of mechanical damage and 'strangles', where the growth of roots has been severely restricted.

61 Downy Mildew *Peronopora farinosa f. sp betae* is a common fungal disease of sugar beet, with beetroots and mangolds occasionally being attacked. Symptoms can often be seen after singling of the plants, where the growing point and central rosette are affected. On young leaves, attacks cause the production of yellowish-green tissues, which become thickened, curled and brittle. In addition to these symptoms, the undersurfaces of the leaves may become covered with a greyish bloom of spores. In severe cases the heart leaves can be destroyed, the damage looking similar to that of Boron Deficiency.

62 *Pythium* is a genus of soil borne fungi belonging to the phycomycetes. Members have well developed mycelia and reproduce sexually by the production of oospores. A large number of species exist, probably the two commonest being *P. debaryanum* and *P. ultimum*, both of which can live as either saprophytes on dead organic matter, or as parasites on living matter. When attacks occur on older beet plants the root tissues are blackened, and necrotic areas appear on the leaves. See Plate 62.

Prevention of the disease can be achieved by good crop management, coupled with the use of seed treated with general fungicides such as thiram and captan.

In addition to the above symptoms, it is possible to find dark brown concentric ringing on beet leaves (See Plate 62F) caused by the fungus *Phoma betae*. However, these symptoms are of little importance and no control measures are required.

63 Beet Rust *Uromyces betae* is a common disease, usually found on soils that have been given excessive applications of nitrogen. Three types of spore are produced during its life cycle, namely, aeciodiospores, uredospores and teleutospores. The first symptoms appear on the under surface of leaves as yellow masses of aeciodiospores, followed later by yellowish-brown uredospores and brown teleutospores. The teleutospores remain

dormant over winter, and produce the primary infection in spring. No control measures are usually necessary for this disease.

Leaf Spots caused by *Ramularia betae* and *Cercospora beticola* produce characteristic light brown or grey spots with brown edges. Damage frequently starts on the lower foliage, but the disease can spread to the younger leaves. Control in seed crops can be obtained by applications of a fungicide such as fentin.

64 Violet Root Rot *Helicobasidium purpureum* is a soil borne fungus which attacks many plants including potatoes, sugar beet, swedes and turnips. It is found in a wide range of soil conditions, but the worst attacks often occur on light alkaline sites. Damage cannot be identified without removing plants from the soil. The symptoms on roots are masses of purple mycelia which can cause soil to stick to the roots, and the formation of areas of brown rot that can spread deeply into the tissues. Dark coloured spots (infection cushions) may be found in the mycelia, and these provide a positive identification feature. Good crop management, weed control, good drainage and maintenance of fertile soils will provide control of this disease.

The fungus *Typhula brassicae* occasionally produces a Root Rot on beet. See Plate 64 for the symptoms.

Grey Mould *Botrytis cinerea* is a common disease of agricultural crops which causes decay of plant tissues. It can usually be identified by the presence of the characteristic grey brown mould. For control measures see notes on Plate 44.

65 Beet Cyst Nematode *Heterodera schachtii*, or Eelworm as it is commonly known, can attack a wide range of agricultural crops such as sugar beet, fodder beet, mangels, turnips, swedes and certain weed species. It is a very serious pest of sugar beet, and because of this the 1977 Beet Cyst Nematode Order was introduced in Britain. This restricts the growing of beet crops on heavily infested soils. The nematode gives rise to patches of stunted, unhealthy plants in fields. On individual plants the outer leaves wilt during hot weather, and eventually turn yellow and die. Stunting of the top root frequently causes a proliferation of lateral roots in which populations of the white, or brown, lemon shaped females build up. As there are no recommended chemicals, adequate crop rotation and a reduction in the movement of soil will provide control of this pest.

Stem and Bulb Eelworm *Ditylenchus dipsaci* attacks a wide range of agricultural and horticultural crops. In beet it produces external cracking of tissues, and when the roots are cut open, a characteristic brown stain may be seen. See Plates 20 and 37.

66 Springtails are primitive wingless insects belonging to the order Collembola. They usually attack plants as secondary pests but in certain conditions they can cause serious damage. Chemical control is achieved by the use of HCH, carbofuran or oxamyl.

Earwigs are nocturnal insects belonging to the order Dermaptera. They are mainly scavengers, but in certain circumstances can cause considerable damage. Earwigs feed on young and immature leaves and may skeletonise the foliage on young growth. Control is provided by general insecticides such as Aldicarb and dimethoate.

Cabbage Thrips *Thrips angusticeps* cause damage to young plant tissues by sucking and scraping at the foliage. Control of Thrips and Capsids can be obtained by chemical sprays of sulphone, dimethoate or fenitrothion.

67 The Peach Potato Aphid *Myzus persicae* overwinters on peach trees in the egg stage; on herbaceous plants and bras-

sicas it overwinters in the immature or adult stages. They migrate from these hosts in early summer, and start to feed on potatoes and other crops. They feed by sucking sap from leaf and stem tissues, thereby causing the loss of large amounts of water and nutrients. Probably more important than the direct damage is their ability to act as virus vectors. Control involves the use of chemicals such as aldicarb, dimethoate, pirimicarb or demephion.

The Black Bean Aphid *Aphis fabae* is a very common pest of agricultural and horticultural crops. It overwinters on the Spindle tree and the Snowball tree, but in late May or early June it migrates to its summer hosts, which may be beans, herbaceous plants and beet. If suitably warm weather prevails the colonies can develop rapidly. In autumn winged females are produced which fly back to their winter hosts. Damage to crops is similar to that caused by the Peach Potato Aphid with large colonies causing severe reduction of growth. In addition to the physical damage, a number of viruses such as Beet Yellows Virus can be transmitted which will reduce growth even more. The control measures are the same as those that apply for the Peach Potato Aphid.

68 A number of beetle species attack the leaves of sugar beet and mangolds. Some of the commonest examples are the Pygmy Beetle *Atomaria linearis*, the Beet Carrion Beetle *Aclypea opaca*, the Beet Flea Beetle *Chaetocnema concinna* and occasionally the Twentyfour Spot Ladybird (*Subcoccinella 24-punctata*). See Plate 69. Pygmy Beetles feed below ground level producing small holes in shoot hypocotyls, and occasionally small round holes in the cotyledons. Carrion Beetles produce irregular holes with blackened edges in the leaves, and in severe attacks the leaves may be completely eaten away. Flea Beetles feed on both surfaces of the leaf, gnawing away the tissues, but leaving a layer of cells intact. In severe cases they can kill plants outright. The Twentyfour Spot Ladybird gnaws the surface of leaves, sometimes producing small irregular shaped holes. Control measures for these insects consist of applying chemicals such as aldicarb, HCH, carbofuran and oxamyl.

69 In addition to the species of beetle mentioned in the notes to Plate 68, damage can also be caused by Cockchafers *Melolontha melolontha*, Garden Chafers *Phyllopertha horticola*, the Striped Tortoise Beetle *Cassida nobilis* and the Clouded Tortoise Beetle *Cassida nebulosa*. The larvae of Cockchafers cause stunting and wilting by feeding on the developing roots, whilst adults can occasionally be seen feeding on the leaves. The various species of Tortoise Beetle are not of serious economic importance in Britain. They eat segments of the cotyledons and give rise to a shot holing of young leaves. Control is achieved by the use of chemicals such as DDT.

70 Sugar beet can be damaged by the larval stages of a number of moths. The species causing the most damage are the Rosy Rustic Moth *Hydroecia micacea*, Cutworms *Agrotis*, *Noctua* and *Euxua spp*, the Cabbage Moth *Mamestra brassicae* and the Mangold Fly *Pegomyia betae*. The Rosy Rustic Moth occasionally attacks plants during May, June and July. It feeds by tunnelling into the developing roots causing wilting and eventual collapse of the plant. Cutworms either feed at night, or less frequently day, just below ground level. The larvae feed on young stems, which may result in stems being severed, poor top growth, wilting and eventually death. Cabbage Moth damage is caused by the larvae feeding on plant leaves. The leaves are largely chewed and often covered with larval excrement. Mangold Fly larvae produce

characteristic 'Leaf Miner' symptoms. Pale coloured, winding tunnels appear in the leaves and eventually join together to produce blisters. The leaves become brown and scorched, and small seedlings can be killed. Control is achieved by applications of chemicals such as DDT azinphos-methyl dimethoate, trichlorphon and aldicarb.

71 Magnesium Deficiency commonly occurs on chalky and light textured soils. It causes chlorosis on the mature leaves of plants. However, leaf margins often remain green and the young tissues rarely show symptoms. Control of this disorder can be achieved by applications of magnesium in the form of Epsom salts, kainit, kieserite or magnesian limestone.

Phosphorus is a major plant nutrient which encourages root development and enhances seedling growth. Deficiency symptoms are uncommon in broad-leaved plants on established arable land, however if it occurs it will lead to poor root development, yellowing of the foliage and reduced yields.

Potassium is another major plant nutrient. A Potassium Deficiency will cause the partial discoloration of leaves. The margins of older leaves are affected first, the affected tissues showing a brown scorching. Deficiencies can be alleviated by applications of potassium fertilisers such as sulphate of potash, potassium nitrate or muriate of potash.

72 Most British soils contain ample supplies of boron, but deficiencies can sometimes occur during periods of drought, after wet winters or where soils contain large quantities of lime. Boron Deficiency in swedes shows as a brown ringing inside the roots, and in severe cases the centre of the root can become rotten. This disorder is given the name of 'Brown Heart' or 'Raan'. It is controlled by applications of sodium tetraborate, the rate depending upon results of soil analysis, remembering that over applications can lead to toxicity.

During periods of drought plants may not be able to take up enough soil moisture for their needs. When this occurs they are said to be under water stress, a condition which will cause tissues to become hard or corky. When water becomes readily available again, the plants absorb large quantities and growth is stimulated, sometimes to such a degree that the toughened tissues expand and crack. This disorder can be prevented by maintaining the soil as close to field capacity as possible.

73 Copper is an essential plant nutrient used in plant respiration and oxidation. Copper Deficiency is not common but symptoms can occasionally be seen on alkaline soils, poor acid sands and fenlands. Symptoms vary, young seedlings being twisted with whitened tissues whilst mature swede plants become generally stunted with small light coloured leaves, which later develop brown patches. Copper sulphate will rectify any deficiency.

Frost on seedlings can cause deformed stunted growth, and even death. On swedes, the leaves may become stunted and malformed with brown edges as a result of frosting. The prevention of Frost Damage to agricultural crops involves the use of the correct variety for a particular site, balanced fertiliser applications to encourage healthy growth and the selection of the correct sowing date.

74 Turnips and swedes can become infected with a number of virus diseases. The three most important are Turnip Mosaic Virus which causes translucency in the main veins, and an interveinal mottling, which is blotched with raised patches of dark green tissue; Turnip Yellow Mosaic Virus which produces vein clearing in the young leaves and small yellow patches on older leaves, and Tur-

nip Crinkle Virus which causes the clearing of veins, distortion, and crinkling of foliage. Severe attacks of this last can lead to rosetting and stunting.

These viruses make sporadic attacks on crops and the control of their vectors (aphids and Flea Beetles) is not usually justified.

75 Frost Damage on swedes produces brown tissue, cracked roots, and allows the entry of secondary diseases. The Swede Midge may cause damage similar to that of frost. For control measures see plate 88.

Black Rot in swede *Xanthomonas campestris* may cause minor damage to crops. It is a seed borne disease causing blackened veins, and browning of both leaf and root tissues. Seed treatment with mercury compounds or hot water treatment give control, but this is usually unnecessary.

76 Bacterial diseases are not as important as fungi, but bacteria can sometimes cause severe damage to crops. They are spread by seed, tubers, wind, water, insects, slugs and snails. Some can live as saprophytes in the soil but others can only live as parasites on plants. Wounds or natural openings in plants allow the entry of bacteria and the resultant damage is caused by the breakdown of tissues. Soft Rot *Pectobacterium carotovorum* is a bacterial disease which affects a wide range of crops such as carrots, swedes and turnips. It enters plants after Slug Damage, Frost Injury or following other diseases and it causes the tissues to disintegrate rapidly, producing a soft smelly rot.

Black Rot *Xanthomonas campestris* is of minor importance in Britain. However, if seedlings are attacked they can be killed outright. Symptoms appear as black staining of veins. The organism is seed borne and infections usually arise from imported seed.

77 Growths that resemble Clubroot may form on cruciferous plants. The formation of hybridisation nodules is an example of this. These growths do not affect plants adversely and control is unnecessary.

Clubroot *Plasmodiophora brassicae* is a common fungal disease of plants belonging to the family Cruciferae. The disease is soil borne and affects the roots producing the common 'Finger and Toe' symptoms. The first damage is seen when plants wilt during warm sunny weather. This is followed by stunting, yellowing and sometimes death. The roots of infected plants become swollen and contorted and when cut through they can have a mottled appearance. The disease spreads by the movement of infected plants, soil and vehicles. Control of the disease can be achieved by general hygiene, correct rotations, the use of lime to raise the soil pH. Soil sterilisation with dazomet, and chemicals such as calomel will also provide control.

78 Downy Mildew *Peronospora parasitica* can cause serious damage to crops, especially during the seedling stages. Leaves become covered with white mycelia and later turn yellow; necrotic brown areas occur, and later the plants are stunted or killed. Control is obtained by applications of dichlofluanid starting immediately after seedling emergence.

White Blister *Cystopus candidus* causes the production of large cells within the hosts tissue. The resulting blistering and distortion is followed by the production of white reproductive bodies. This disease is frequently found in association with Downy Mildew.

White Spot in swede *Cercosporella brassicae* causes leaves to beome yellow and flecked with light spots encircled with darker brown margins.

Mildew *Erysiphe polygoni* attacks a range of crops making the leaves white with powdery mycelia. Yield may be slightly reduced, but as with White Blister and White Spot, control is usually unnecessary.

79. Leaf Spots of brassicas can be caused by *Alternaria brassicae* and *A. brassicola*. With the former species, attacks result in circular brown spots on the leaves whereas the latter species produces spots on leaves, stems and pods. Both can be controlled by the use of thiram seed soaking treatments.

Dry Rot Canker *Phoma lingam* is a seed borne disease causing brown lesions on seedling cotyledons. Brown depressed cankers may be seen on older roots. The disease is controlled by the method used for leaf spots.

80 Rape and other cruciferous plants can be damaged by the Stem and Bulb Eelworm *Ditylenchus dipsaci*. This pest causes the distortion of foliage, galling of individual leaves, the death of growing points and a general stunting of growth. Control may be difficult, but aldicarb, oxamyl and thionazin are all effective. *Eurydema* and *Calocoris spp* may attack swedes in a similar manner, however they can be effectively controlled by dimethoate, formothion and malathion. See also notes to Plate 66.

81 In addition to damage caused by the species seen on Plate 80, cruciferous crops can be infested with Cabbage Thrips *Thrips angusticeps* and Cabbage Aphid *Brevicoryne brassicae*.

Thrips are minute insects that can be seen in large numbers during the autumn; in winter they hibernate in the soil, and emerge in spring to attack young seedlings. Silver mottling of foliage and a general reduction in growth are the symptoms of an attack by Thrips. For control see below.

Cabbage Aphid is a serious pest of brassicas. The damage is most severe on young plants, where small bleached areas occur on the leaves. The damaged foliage later becomes yellow and distorted, and in serious attacks whole plants may be severely stunted. Control is achieved by the use of insecticides such as demeton-S-methyl and dimethoate.

82 The Turnip Flea Beetle *Phyllotreta nemorum* attacks a wide range of cruciferous plants including turnip, swede, radish, mustard and cabbage. The adult beetles eat small holes into the stems and leaves of seedlings. This is most serious in dry springs when the plants are under water stress. The beetles are small and difficult to see, but in later attacks they may be seen in the growing rosettes of the plants. Control can be obtained by seed treatments (HCH) and chemical sprays of DDT or carbaryl.

83 Blossom Beetles *Meligethes spp* are serious pests of brassica seed crops. They are found on plants just before flowering, where they bite out tiny slits in the base of flower buds. These flowers later become shrivelled and die. Control measures consist of applying insecticides such as azinphos methyl, Gamma-HCH or malathion immediately prior to flowering. For good control at least two applications are necessary.

The larvae of Cabbage Stem Flea Beetles *Psylliodes chrysocephala* burrow into the leaf stalks and stems of brassica plants. During heavy infestations, the plants may lose their leaves, collapse and die. Control can be obtained with the insecticides used for Blossom Beetles.

84 Cruciferous crops can be seriously damaged by a number of Weevils, the commonest of which is the Cabbage Stem Weevil *Ceutorrhynchus quadridens*. The larvae of the Weevil bore into stems and feed voraciously. Other common species include the Turnip Seed Weevil *Ceutorrhynchus assimilis* whose larvae feed on developing seed pods, and the Cab-

bage Gall Weevil *Ceutorrhynchus pleurostigma* whose larvae live and feed in root galls.

In general, control of these pests can be achieved by using the chemicals applied to control Blossom Beetles. See notes to Plate 83. A reduction in numbers of the Cabbage Stem Weevil can be obtained by using a Gamma-HCH/thiram seed dressing.

85 The Turnip Sawfly *Athalia rosae* is an occasional pest of turnips in Britain. The adults are yellow and approximately 6–9mm in length. They are on the wing from the beginning of May when they lay their eggs on the margin of turnip leaves. Damage is caused by the larvae feeding on the leaves, and in severe attacks the foliage can become skeletonised. Where necessary, the larvae can be killed by using contact insecticides.

The larvae of the Chrysomelid Beetle *Colaphellus sophiae* can produce leaf damage similar to that caused by Turnip Sawfly. Although its body structure differs from the Sawfly, the Chrysomelid Beetle is controlled by the same method.

86 Larvae of the Diamond Back Moth *Plutella xylostella* sometimes build up in large numbers, and can severely damage members of Cruciferae. The Moths appear in May–June to lay eggs on cabbages and related species. The eggs hatch to produce pale green caterpillars which feed in a web on the undersurface of leaves. Most damage is caused to interveinal areas, the larvae often rejecting the upper surfaces and veins. Damaged leaves appear to be translucent, but later become brown and shrivelled. After about three weeks the larvae are fully grown. They then pupate, and emerge as adults after approximately two weeks. Two generations each year are common, but in mild seasons three may occur. Control can be achieved by spraying the undersurface of leaves with chemicals such as DDT, azinphos-methyl, iodofenphos, triazophos and trichlorphon.

Injury to plants caused by Cutworms is also illustrated on this plate. For control see Plate 70.

87 The Cabbage White Butterfly *Pieris brassicae* and the Small Cabbage White Butterfly *Pieris rapae* attack members of the Cruciferae. Damage is caused by the larvae which eat plant leaves. Where Large Whites are present foliage can be ravened, the plants soon becoming skeletonised. Damage from the Small White is less dramatic, but in certain circumstances it can be of greater economic importance. The life cycles of the two species are similar, except that the Large White emerges a little later, and instead of laying solitary eggs, it lays large batches on the undersurface of leaves. In both cases there is more than one generation each year, the second generation usually causing the worse damage. Control measures consist of timely application of insecticides such as DDT, mevinphos and trichlorphon, but some control can be achieved by natural enemies. The commonest example of this is the Braconid Wasp *Apanteles glomeratus* whose larvae feed inside the bodies of the caterpillars.

88 Swede Midge *Contarinia nasturtii* can be an important pest in areas where swedes are grown for seed. Most damage is caused during June when the Midge feeds in the growing tips of swede plants. This causes the death of main shoots and the growth of many competing shoots. Leaves and inflorescences are also deformed. Good crop management and adequate rotation will provide control.

89 Leatherjackets are the larval stage of Crane Flies *Tipula spp*. They feed by eating the roots and underground stems of

many plants. They feed just beneath soil level, but during warm, damp nights they can feed on the surface, when they cause damage similar to that of Cutworms. Control of this pest can be obtained by insecticides such as HCH, DDT or fenitrothion, coupled with correct rotational and cultural practices.

Various species of Root Fly attack agricultural crops. *Erioischia brassicae* and *E. floralis* can cause serious damage to members of Cruciferae. The larvae of both species feed on roots causing the death of seedlings and severe checks in the growth of older plants. There are usually two generations each year, but in mild seasons a third can occur. Overwintering takes place in the pupal stage, the adults emerging in the spring. Control can be achieved by using insecticides such as carbofuran on swedes and turnips, or chlorfenvinphos and diazinon on brassicas. A method of assessing Root Fly populations using pheromones (natural chemical stimulants), has recently been developed at the National Vegetable Research Station, and this should be an aid in the control of these pests.

90 Potatoes require large amounts of potassium for healthy growth, and if any deficiency occurs they quickly show symptoms. These appear as the yellowing of leaf margins on the older leaves, followed by brown scorching and localised death of the leaf. The production of dark green leaves, which later become purple or bronze, and have the tendency to crack and die, is also symptomatic. Deficiencies can be alleviated by applications of potassic fertilisers.

Physical damage to potatoes is seen in the yellowing of affected leaves followed by a browning of the leaf margins. In severe cases the leaves can crack and shrivel.

Phosphorus, like potassium, is an essential plant nutrient required in large amounts. Small sized plants, with slow growth and dull coloured leaves indicate a deficiency of phosphorus. Confirmation of the deficiency can be made by soil analysis.

91 Plate 91 shows Phosphate Deficiency in potatoes grown on reclaimed Calluna heath, Denmark; the site was ploughed and marled in 1939. The area in the foreground has never received phosphate fertiliser, whereas the plot in the background was given 1000 kg/ha superphosphate in 1940. The photograph was taken in July 1947. For the symptoms, compare this with Plate 90 D.

Plate B shows Phosphorus Deficiency in swedes. The experiment was the same as above, the photograph being taken in July 1944. For the symptoms, compare this with Plate 71 B.

92 Manganese is a minor nutrient which can become deficient in light, alkaline soils rich in organic matter. A deficiency will cause chlorosis and in some varieties dark spots on the leaflets on either side of the main veins. Applications of manganese sulphate will rectify this deficiency.

Magnesium Deficiency discolours the older leaves and is followed by yellowing between the veins. The leaf margins often remain unaffected. In severe cases the leaves become progressively discoloured turning yellow, then brown, before dying. Deficiencies can be alleviated by applications of Epsom salts either as foliar sprays or soil applications.

Nitrogen Deficiency renders plants stunted, with pale yellow-green leaves. However, care should be taken in its identification because the symptoms may be similar to Magnesium Deficiency.

Potatoes are very susceptible to the over application of borax, which will cause yellowed leaves with large black spots between the main veins (compare this with the Manganese Deficiency above).

93 During periods of hot dry weather plants may have difficulty in obtaining enough water to sustain healthy growth. When this occurs they are said to be under water stress. This results in a severe reduction in growth and a hardening of the tissues. When water becomes readily available again, rapid growth can take place, which may cause potato tubers to crack, producing secondary growth and internal cavities. The application of sufficient quantities of water before or during dry spells will overcome these problems.

94 Net Necrosis is sometimes a symptom of Potato Leaf Roll Virus. When cut in cross section, young tubers exhibit spots or fine streaks in two concentric rings, and when cut obliquely the streaks make an irregular network. The symptoms usually become more intense after lifting, and infected tubers should not be used as seed.

The cause of Rust Spot may be a virus but this has not yet been confirmed definitely. The disease gives rise to distinct brownish blotches on cut surfaces; these do not appear to increase during storage. No means of prevention are known.

Spraing is thought to be caused by a virus (Tobacco Rattle). It causes reddish brown lesions on cut tissues, and in cross section it makes a curved, wavy or arc-like pattern. No means of control are known.

Plates D–I show the damage caused to tubers after varying degrees of frost. J illustrates the proliferation of tubers after planting, and K shows the production of spindly shoots.

95 Rugose Mosaic is a virus disease spread by aphids. It causes a reduction in yield, and if infected tubers are planted they reproduce the disease. In the first year of infection the commonest system is the production of dark brown streaks running along the veins on the lower surface of the leaves. The second year of infection leads to dwarf growth and roughened yellow leaves. Later in the second year stems become weakened and lie on the ground. Control can be achieved by roguing out infected second year plants, using good quality seed, and controlling aphid vectors with insecticidal sprays. However in some cases aphid control has been shown to increase the spread of the disease.

Potato Aucuba Mosaic Virus symptoms differ between varieties, however they often appear as conspicuous bright yellow spots or patches scattered over the dark green leaf surfaces. Control measures consist of planting certified seed and controlling aphids with insecticides.

96 Leaf Roll Virus is an aphid transmitted disease of potatoes causing large reductions in yield. The tubers show no visible external symptoms, but in certain varieties a condition known as Net Necrosis can be seen (see Plate 94). No visible symptoms appear in the first year unless infection takes place early in the season. When this occurs the leaflets of the upper leaves become purple and start to roll (Primary Leaf Roll). In the second year, when infected tubers are planted out the symptoms are more pronounced. They appear a few weeks after emergence when the lower leaflets curl inwards. Later in the season the whole is affected in this way. In severe infections the lower leaves can turn brown, and the leaves rattle when shaken. To control this virus good quality seed should always be used and aphids should be controlled with insecticides. If seed crops are involved they should be grown in isolation and the infected second year plants rogued out.

97 Potato Virus A can produce a faint mottling of the foliage on potatoes, however, when in association with other

viruses such as Potato Virus X or Y, it can cause the disorder known as Crinkle. Plants infected with Crinkle become stunted with pale green, puckered leaves and small tubers.

Mild Mosaic Virus may be carried in some plants without exhibiting visible symptoms, whilst other plants develop a faint mottling with two distinct shades of green on the leaves. The disease is spread by contact, and aphids are not involved in its transmission.

98 Leaf Drop Streak Virus (Virus Y) produces a blotchy mottling on the foliage of potatoes. It spreads from the veins and affects the topmost leaves only. In severe cases, fine dark brown discolorations can be found running along the veins and down the leaf stalks. Leaves eventually become necrotic and wither but do not fall from the plant. The disease can be prevented by growing certified seed.

99 Blackleg of potatoes *Pectobacterium carotovorum var. atrosepticum* is a widespread disease in this country. The disease may appear in early summer, however attacks are more often seen in June just before the haulms meet across the rows. The type of damage varies with the stage of plant growth. Young seedlings may be killed before, or just after emergence. Plants infected in mid-summer appear stunted with light green or yellowish foliage. The upper leaves become stiff and erect, whilst the leaf margins roll inwards. A black stem rot may be found at, or below, ground level and the vascular strands are often stained. Late infections can cause the leaves to turn brown and die, whilst the stem bases show a black, wet, soft rot. Control measures consist of roguing out infected plants, and buying in clean certified seed.

100 With the gradual reduction of available farm labour and the inevitable increase in mechanisation, the incidence of mechanical or physical damage increases. This, coupled with incorrect storage techniques, can lead to large losses of potato tubers due to the development of internal rots. Bacterial Soft Rot caused by *Pectobacterium carotovorum* or other bacteria frequently occurs as a secondary agent following damage caused by suffocation, frosts, slugs or mechanical injury. The disease invades damaged fleshy tissues on a range of plants, causing a rapid disintegration of cells with the subsequent development of a soft smelly rot. The correct storage of crops and the prevention of physical and mechanical damage will control these bacterial rots.

101 Wart Disease caused by the fungus *Synchytrium endobioticum* has declined in importance in recent years. This is because immune varieties have been developed and stringent statutory regulations have been enforced in Britain by the Ministry of Agriculture, Fisheries and Food. Damage is usually restricted to tubers, the foliage not being affected to any extent. However in some instances warty outgrowths can form on the basal stems and leaves. Infected tubers can be covered with warts, or there may be just localised clusters, and in some cases undamaged tubers can be found on badly infected plants. The disease spreads in the soil by spores which are resistant to adverse conditions. Control measures involve growing immune varieties and following the regulations laid down.

Powdery Scab *Spongospora subterranea* is occasionally mistaken for Common Scab. However, its spores tend to be more circular in shape than those of common scab. The scabs which may be solitary or in clusters, eventually rupture to give a dark mass of brown spores. It is these spores which rest in the soil and spread the disease. In some circumstances the disease can cause a malformation of the tubers giving rise to ugly cankers. The wide rotation of crops on well drained land, coupled with the legislation preventing badly infected tubers being sold for seed provides control of the disease.

102 Potato Blight *Phytophthora infestans* is the most serious disease of main crop potatoes in Britain. The first symptom is an occasional spotting of the leaves and to a lesser extent the stems. The spots are dark, with pale coloured margins and on the underside of the leaf the spots are covered with a white fungal web full of spores. The spores are washed into the soil by rainfall and this gives rise to the infection of the tubers. Once fully established the disease spreads rapidly and plants can lose most of their foliage within a fortnight. Control can be achieved by spraying fungicides, such as captafol, mancozeb, nabam and zineb at regular intervals. A Ministry of Agriculture forecasting scheme is in operation throughout Britain based on Beaumont periods, which gives growers warning of when attacks of the disease are likely.

Damage to potatoes similar to that caused by Blight, may occur as a result of Leaf Spot *Cercospora concors* which produces small brownish-black spots, and Early Blight *Alternaria solani* which produces brown or black angular spots on leaves or dark spots on stems. If the identification of these diseases proves to be difficult it is best to consult a local fieldsman of the Agricultural Development Advisory Service.

103 Potato tubers are susceptible to damage from a number of causes. Late Blight *Phytophthora infestans* can produce dark, sunken, irregular shaped areas, and when the tubers are cut open a reddish-brown rot is seen. In wet conditions the tubers can be invaded by secondary organisms causing wet rots, although in other cases they can remain firm and only exhibit sunken lesions.

Damage can also be produced by mechanical injury during harvest, or by organisms such as *Fusarium spp*, and *Alternaria solani*. The most serious of these is the dry rot caused by *Fusarium coeruleum*. This organism develops quickly in high humidities and temperatures around 16°C. In these conditions the symptoms become visible about three to four weeks after infection, although in lower temperatures development is much slower. The symptoms appear as small brown patches on the tubers which become sunken and showing concentric wrinkles. Later, the skins crack and mycelia covered with white or pink spores are produced. The control measures involve growing resistant varieties, preventing mechanical damage and chemical disinfection using tecnazine.

104 Damage to potatoes by the soil borne fungus *Corticium solani* can vary depending on which part of the plant is attacked. The infected tubers exhibit small, very dark encrustations, resembling pieces of coal, which give it the name Black Scurf. These superficial crusts are the resting stage of the fungus. Other symptoms can be seen on the young stems where cankers are formed which can kill plants and lead to bare patches in the field. It is sometimes possible to find a cottony collar at the base of stems; this is the spore bearing stage of the fungus. Prevention of this disease can be obtained by planting healthy, chitted seed and treating infected seed tubers with mercury compounds.

105 In addition to the disorders seen on Plates 93, 94, 99, 100, 101 and 103, potato tubers can be damaged by Common Scab *Streptomyces scabies* where the

symptoms appear as loose corky growth near the point where the infection took place; Powdery Scab *Spongospora subterranea* producing spots similar to Common Scab, or cankerous growths which are sometimes confused with Wart Disease; Silver Scurf *Spondylocladium atrovirens* causing silvery patches on the skin of tubers, and Skin Spot *Oospora pustulans* which stimulates the production of small pimple-like spots on the skin of tubers. These spots are often dark and sunken, with raised centres.

Generalised control measures for these diseases consist of practising adequate rotations, growing clean healthy seed, and following correct manural practices. However, where outbreaks occur in Britain specialist advice can be obtained from the Ministry of Agriculture Fisheries and Food.

106 A number of Nematodes or Eelworms attack the roots of potato plants, some of which are serious problems in potato growing areas. Where crops are grown regularly large Nematode populations can build up very quickly. The Potato Root Nematode *Heterodera rostochiensis* lives in the soil in cysts which are easily seen and identified. When young, the cysts are shiny white but with age they turn brown. Another species which attacks potato crops is the Potato Tuber Nematode *Ditylenchus destructor*. This species lives in tubers causing shrivelling and browning of the tissues. The control of these pests is regulated by government legislation and involves careful cultivation, crop rotation, resistant varieties and the use of chemicals.

107 Colorado Beetle *Leptinotarsa decemlineata* is a very serious pest of potatoes, but it has not been allowed to become established in Britain. The adults are approximately 12 mm long and covered with black and yellow stripes. The larvae are orange/brown when young turning red with age. Damage starts when the adults feed on the leaves, and this is continued by the larvae. Plants can be completely defoliated, and covered with dirty excrement. Government legislation requires farmers to report suspected outbreaks, and also prohibits the keeping of live specimens. Precautionary control measures have been carried out in areas where the Beetle has been suspected of being imported.

108 Potatoes are susceptible to attack from a number of insect pests; some of the commonest are Wireworms, Capsids, Rosy Rustic Moths and Cockchafers. The majority of damage is caused by larval feeding, although in certain circumstances the adults can be troublesome. The symptoms of attack range from burrows in stems, roots and tubers, to the destruction or malformation of foliage. More specific details of the symptoms and damage are given in notes on Plates 22, 66, 69, 70 and 80. Cultural control measures for the insects vary with the species concerned, but the use of insecticides such as DDT and phorate give good control.

109 *Sclerotinia sclerotiorum* produces a soft rot in stored crops such as carrots and potatoes. It results in rotted tissues covered with a white fluffy mycelium containing the overwintering sclerotia. Control can be achieved by burning all debris, preventing physical damage, and storing only sound roots.

Carrots may be affected by Violet Root Rot, and a black rot caused by the fungus *Stemphyllium radicinum*, the symptoms of which appear as black sunken lesions usually at the crown, but in serious attacks the whole root can be affected. For control measures see above.

Millipedes *Blaniulus guttulatus* damage crops by gnawing roots and allowing the entry of secondary organisms. If they become serious enough to warrant control, chemicals such as aldicarb, DDT or HCH are effective.

110 A number of insects feed on carrots. Seed moths *Depressaria spp* live on inflorescences or leaves; Cutworms and Carrot Flies feed on roots, and Aphids suck plant sap. Carrot Fly *Psila rosae* is a very serious pest causing severe damage to crops. The larvae feed by tunnelling in carrot roots and in heavy attacks the foliage becomes reddish and later shrivels. The leaf symptoms can be similar to those caused by the Willow Carrot Aphid *Caveriella aegopodii*. Insecticides such as phorate, chlorfenvinphos, disulfoton and triazophos will give control.

The Willow Carrot Aphid can cause considerable loss of yield in carrot crops. The Aphid distorts foliage, making it reddish or yellow in colour, and in severe infestations the foliage is stunted and killed. This Aphid is a vector of the Carrot Motley Dwarf Virus which in itself can seriously reduce yields. The control consists of insecticidal applications using chemicals such as demephion, dimethoate, formothion or phorate.

111 Manganese is a nutrient that plants require in small amounts. It may be deficient in light land which is high in organic matter and contains an excess of lime. Deficiency symptoms in flax appear as the formation of light green leaves which later turn brown and wither. Applications of manganese sulphate to the soil will rectify the deficiency.

Seedling Blight *Colletotrichum lini* can be a destructive disease of flax seedlings. The symptoms appear as brown lesions on the young leaves and stems. Flax plants may recover from mild infections but severe attacks will kill plants. This is a seed borne disease which can be controlled by treating the seed with fungicides such as thiram, growing resistant varieties and ensuring a high standard of hygiene.

Browning and Stem Break caused by *Polyspora lini* is a seed borne disease which causes plants to be browned just before harvest. The browning is a result of the formation of brown lesions on the stems, leaves and flower heads. In severe attacks large patches of plants can become infected. When young plants are attacked they may bend over or even snap off. The control involves seed treatment with thiram or mercury-based compounds.

112 Flax Rust *Melampsora lini* is an obligate parasite capable of seriously reducing plant growth. The symptoms appear in the form of irregular yellow structures on the young leaves, and later in the season brownish-black spores can be produced on stems and leaves. The best method of control is to grow resistant varieties.

Pasmo Disease of flax *Septoria linicola* affects the cotyledons during germination although the damage becomes apparent only just prior to flowering. The disease causes dark circular spots encircled with distinctly lighter margins on the leaves, and dark lesions on the stems, leaves and bolls. The use of resistant varieties provides an effective control.

113 Barley Yellow Dwarf Virus can be an important disease of cereals such as barley, oats and wheat. It also occurs on a variety of grass species. The virus first shows as a yellowing of the leaf tips, and later this extends down the leaf. The affected plants become stunted, ears may be blasted and a reduction in yield results. It is important to remember that the symptoms of this virus can be similar to those of other agencies such as Nitrogen Deficiency. Since this virus is transmitted by aphids applications of insecticides will effectively reduce its spread.

114 In addition to Yellow Dwarf Virus, barley can show leaf damage caused by Barley Stripe Mosaic Virus. However, control measures for this disease are usually unwarranted.

Leaf Stripe of barley *Pyrenophora graminea* produces a pale striping on the emerging leaves, and in severe attacks seedlings can be killed. Plants which survive initial attacks have pale-green striped foliage which later turns yellow and cracks. Control can be achieved by adequate crop rotations and carboxin-thiram seed treatments.

Grasses can be affected by a number of virus diseases. Cocksfoot Mottle Virus, Bearded Couch Grass Mosaic Virus and the Rye Grass Mosaic Virus are examples. The last can cause severe damage especially in the southern half of England. Symptoms appear as mottling and streaking of foliage, or the formation of dark brown areas on individual leaves. This virus is spread by mites and at the present time no economic control measures are recommended.

115 Wheat Rust *Puccinia striiformis* can cause considerable damage to cereal crops. Lines of yellow pustules appear on the leaves running the length of the veins. Later in the season these pustules may turn black. Control can be obtained by growing resistant varieties or by using protectant sprays of benodanil or tridemorph with polyram.

Glume Blotch *Leptosphaeria nodorum* produces a variety of symptoms depending upon which part of the plant is affected. Glumes may be covered with irregular brown spots, ears may be blackened with secondary infections of other diseases, leaves may exhibit pale yellow spots, and stems can show browning of tissues near the nodes. The bendimidazole fungicides will provide control.

Mildew *Erysiphe graminis* can be a very serious problem in barley crops. White or brown masses of mycelia build up on the ears, lower leaves, and as the disease develops the mycelial patches turn brown and spores are produced. Control involves the use of ethirimol seed treatments and chemical sprays of ditalimfos, tridemorph or triforine.

116 A number of aphid species attack cereal crops in Britain. The Grain Aphid *Sitobion avenae*, the Rose Grain Aphid *Metapolophium dirhodum* and the Bird Cherry Aphid *Rhopalosiphum padi* are common examples.

The main damage to cereal crops is caused by the aphids sucking sap from the plant tissues, but secondary damage can be produced by their action of transmitting virus diseases. See the notes to Plate 113 on the Barley Yellow Dwarf Virus. The symptoms of direct attack vary with the individual species concerned, but in general plants become pale and stunted. Chemical control may be applied with either spray or granule formulations. As a general recommendation, treatment is warranted when five aphids are found on each ear at the beginning of flowering. When infestations occur during other growth stages it is advisable to consult a local Agricultural Development Advisory Service fieldsman. The following chemicals are commonly used in aphid control: demeton-S-methyl, dimethoate, phosalone and pirimicarb.

117 A number of fly species can cause damage to cereals. The larvae of the Wheat Leaf Mining Fly *Hydrellia griseola* produces light yellow coloured areas on leaves. The larvae of the Gout Fly *Chlorops pumilionis* feed in the shoots of barley, wheat and rye. Young damaged shoots appear swollen and in the later stages they can rot and shrivel away. The larvae of various species of Sawfly *Dolerus spp* can feed on the ears of barley and cause severe damage to leaves. In many cases the control of these pests is un-

warranted but they can usually be reduced by following good cultural practices.

118 Grasses are occasionally attacked by species of *Pyrenophora* and *Rhynchosporium*. They cause yellowing and blotching, or streaking and spotting of the leaves.

Crown Rust *Puccinia coronata* can be a serious problem in rye grass swards. The symptoms appear as yellowing of foliage, coupled with the production of black and orange spores. Control can be obtained by fungicidal sprays of maneb and nickel sulphate; some degree of resistance has been found in some cultivars.

Mildew *Erysiphe graminis* can cause damage to meadow grass. The white powdery mycelium on the upper surface of the leaves is most damaging after dry spells. Control can be achieved by balanced applications of nitrogen and potassium fertilisers.

Leaf Spot *Mastigosporium rubricosum* produces dark brown spots on the leaves of bearded couch grass. For symptoms see notes to Plate 118.

119 In addition to the troubles seen on Plates 90–108, potatoes can be damaged by Bacterial Ring Spot *Corynebacterium sepedonicum*, Potato Gangrene *Phoma solanicola F. Foveata* and the over application of chemicals such as reglone. Of the three, Gangrene causes the most serious damage. It is mainly a disease of stored potatoes, which produces small dark, oval or round, sunken patches. In severe cases wrinkled lesions can be produced over a deep rot. Control can be obtained by using the fungicides 2-aminobutane and thiabendazole or by dipping newly harvested tubers in certain organomercury compounds such as methoxyethyl mercurous chloride.

For Common Scab see Plate 105 and for Potato Tuber Nematode see Plate 106.

120 Other than the troubles seen on plates 109 and 110 carrots be damaged by the Root Knot Nematode *Meloidogyne hapla* which causes stunting and galling of growth; Boron Deficiency, appearing as orange-red foliage and a reduction in growth; Cavity Spot, forming sunken areas in the roots, but without the presence of mycelia; Grey Mould *Botrytis cinerea* which produces a greyish-white mass of mycelia over stored crops, and Crater Rot *Rhizoctonia carotae* that produces white sunken areas on the roots.

INDEX OF ENGLISH NAMES

A
Anthracnose 40, 142
Aucuba mosaic virus 95, 155

B
Bacterial ring spot 119, 161
 soft rot 100, 120, 156
Barley stripe mosaic virus 114, 160
 yellow dwarf virus 113, 159, 160
Bearded couch grass mosaic virus 114, 160
Beet carrion beetle 68, 149,
 cyst nematode 65, 148
 flea beetle 68, 149
 mild yellowing virus, 146
 mosaic virus 56, 145, 146
 rust 63, 147
 yellows virus 146, 149
Bibionid flies 24, 139
Bird cherry aphid 116, 160
Bird damage 28, 139
Black bean aphid 48, 67, 144, 149
 blotch 34
 fly *see* Black bean aphid
 globular springtail 66
 leg 60, 99, 156
 mould 10, 56, 135
 rot 75, 76, 109, 151
 rust 15, 137
 scurf 104, 157
 stem fungus 41, 142
Blossom beetle 83, 152, 153
Borax injury 8, 92, 135, 154
Boron deficiency 8, 40, 50, 51, 72, 120, 134, 142, 145, 147, 150, 161
Braconid wasp 87, 153
Broad bean stain 142
Broad bean true mosaic 142
Brown foot rot 12
 heart 72, 150
 rust 15, 31, 137, 140
Browning and stem break 111, 159
Bunt 17, 137

C
Cabbage aphid 81, 152
 gall weevil 84, 152
 maggot 75
 moth 70, 149
 root fly 89, 154
 stem flea beetle 83, 152
 stem weevil 84, 152

 thrips 66, 81, 148, 152
 white butterfly 87, 153
Capsids 148
Carrot fly 110, 159
 motley dwarf virus 159
Cavity spot 120, 161
Cereal cyst eelworm 20, 137
 leaf beetle 26, 30, 139
 mildew 13, 29, 115, 118, 136, 140, 160, 161
 root eelworm 30, 140 *see also* Cereal cyst eelworm
Choke 29
Chrysomelid beetle 85, 153
Click beetle 22, 138
Clouded tortoise beetle 69, 149
Clover leaf weevil 38, 142
 rot 35, 41, 141, 142
 seed weevil 38, 142
 sickness 37, 141
Clubroot 77, 151
Cockchafer 22, 69, 108, 149, 158
Cocksfoot mottle virus 160
Colorado beetle 107, 158
Common earwig 36, 66, 141, 148
 green capsid 108
 rustic moth 23, 138
 scab 59, 105, 119, 147, 156, 157
Copper deficiency 6, 73, 134, 150
Covered smut 18
Crane fly larvae 24, 89, 139, 153
Crater rot 120, 161
Crinkle 97, 156
Crown gall 58, 146
 rust 15, 31, 118, 137, 140, 161
Cutworm 70, 86, 110, 149, 153, 154, 159
Cyanamide scorch 4, 133

D
Dark leaf spot 79
 spot 79
Diamond back moth 86, 153
Downy mildew 41, 61, 78, 142, 147, 151
Drought damage 6, 134
Dry rot 103, 157
Dry rot canker 79, 152

E
Ear blight 12, 136
 smut 30
Early blight 102, 103, 156
Earwig *see* Common earwig

Eelworm 20, 106, 138, 148, 158
Ergot 13, 29, 136, 140
Eyespot 11, 135, 136

F
Finger and toe 151
Flax rust 112, 159
 seedling blight 111, 159
Flounced rustic moth 32, 140
Foot rot 12, 44, 136
Frit fly 25, 139
Frost damage 7, 29, 39, 53, 54, 73, 75, 92, 94, 109, 139, 142, 145, 150, 151

G
Gall midge 24, 31, 41, 43, 88, 139, 140, 143
Garden chafer 69, 149
Girdle scab 59, 147
Glume blotch 115, 160
Gout fly 117, 160
Grain aphid 21, 116, 160
Grey field slug 28, 36, 141
 mould 44, 47, 64, 120, 143, 144, 148, 161

H
Hail damage 54, 145
Halo blight 9, 13, 135
Heart rot 50, 145
Hessian fly 32, 140
Honeydew fungus *see* Sooty mould

I
Ion injury 8
Iron deficiency 8, 135

L
Large cabbage white butterfly 87, 153
Late blight 103, 157
Leaf and pod spot 44, 143
 beetle 30
 blight 103, 157
 blotch 14, 47, 118, 136, 144
 drop streak virus 98, 156
 fleck 29, 140
 midge 43, 143
 miner 26, 139, 150
 roll virus 96, 155
 spot 9, 13, 34, 41, 46, 62, 63, 79, 102, 118, 135, 136, 141, 142, 144, 148, 152, 157, 161
 streak 118
 stripe 114, 160

Leatherjacket *see* Crane fly larvae
Lightning damage 54, 145
Loose smut 17, 18, 137
Lucerne flower gall midge 43, 143
   leaf gall midge 43, 143
   leaf spot 41, 142
   leaf weevil 42, 143
Lupin seed fly 48, 144

M

Magnesium deficiency 4, 49, 71, 92, 134, 144, 150, 154
Manganese deficiency 4, 5, 9, 52, 57, 70, 92, 111, 134, 135, 145, 149, 154, 159
Mangold fly 70, 149
Meadow foxtail gall midge 31
   grass rust 31, 140
   plant bug 32, 140
Mild mosaic virus 97, 156
Mildew 13, 29, 35, 41, 44, 61, 78, 115, 118, 136, 140–3, 147, 151, 160, 161
Millipedes 109, 158
Mouse damage 28, 139

N

Neck rot 75
Nematode 106, 158 *see also* Eelworm
Net blotch 118
   necrosis 94, 155
Nitrogen deficiency 1, 49, 92, 133, 138, 144, 154, 159

O

Oat cap 12, 136
   spiral mite 27, 139

P

Pasmo disease 112, 159
Pea and bean weevil 39, 46, 142, 144
   aphid 45, 143
   beetle 45
   moth 46, 144
   noctuid moth 43
   rust 45, 143
   thrip 45, 143
Peach potato aphid 67, 146, 148, 149
Phosphorus deficiency 2, 71, 90, 91, 133, 150, 154
Pit rot 103
Potassium deficiency 3, 33, 39, 44, 49, 71, 90, 133, 135, 140, 142–4, 150, 154
Potato aucuba mosaic virus 95, 155

Potato blight 102, 157
   capsid bug 66, 80
   gangrene 119, 161
   leaf roll virus 96, 155
   root eelworm 106, 158
   root nematode *see* Potato root eelworm
   tuber eelworm 106, 119
   tuber nematode *see* Potato tuber eelworm
   virus A 97, 155
   virus X 97, 156
   virus Y 97, 156
Powdery scab 101, 105, 156, 158
Primary leaf roll 96, 155
Pygmy mangold beetle 68, 149

R

Raan 150
Red oats 1
Ribbon leaf 8
Ring spot 34
Root fly 89, 154
   knot eelworm 120, 161
   knot nematode *see* Root knot eelworm
   rot 48, 58, 60, 64, 144, 147, 148
   weevil 43, 143
Rose grain aphid 116, 160
Rosy rustic moth 70, 108, 149, 158
Rugose mosaic 95, 155
Rust 15, 31, 36, 45, 63, 112, 136–7, 140, 141, 143, 147, 159
   spot 94, 155
Rye grass mosaic virus 114, 160

S

Saddle gall midge 19, 137
Sand weevil 84
Sawfly 117, 160
Scorch 34, 141
Sea salt injury 8, 135
Seed moth 110, 159
Seedling blight 111, 159
Septoria leaf spot 9, 135
Sharp eyespot 11, 136
Silver scurf 105, 158
Skin spot 105, 158
Slug damage 28, 139, 151
Small cabbage white butterfly 87, 153
Smuts 17, 18, 30, 137, 140
Snow mould 12, 136
Sodium chlorate injury 8, 135
Soft rot 76, 151

Sooty mould 67
Speckled yellows 52, 145
Spraing 94, 155
Springtail 66, 148
Stem and bulb eelworm 20, 37, 42, 65, 80, 138, 143, 148, 152
Strangles 60, 147
Stripe smut 17, 137
Striped tortoise beetle 69, 149
Sugar beet eelworm 65, 148
Swede midge 75, 88, 151, 153
   mosaic virus 74

T

Tail rot 58, 147
Take all 10, 135
Tarsonemid mite 32, 140
Thrips 21, 66, 81, 138, 148, 152
Thunderflies *see* Thrips
Timothy fly 30, 140
   tortrix moth 32, 140
Tobacco rattle 94, 155
Turnip crinkle virus 150
   flea beetle 82, 152
   mosaic virus 74, 150
   pod midge 84
   root fly 89, 154
   sawfly 85, 153
   seed weevil 84, 152
   yellow mosaic virus 74, 150
Twentyfour spot ladybird 68, 149

V

Verticillium wilt disease 40, 142
Violet root rot 64, 104, 109, 148, 158
Virus yellows 55–7, 146

W

Wart disease 101, 156, 158
Water stress 93, 155
Weevils 38, 84, 142, 152
Wheat bulb fly 26, 139
   flea beetle 22, 138
   leaf mining fly 117, 160
   midge 24, 139
   rust 15, 115, 137, 160
   stem sawfly 23, 138
White blister 78, 151
   clover seed weevil 38
   spot 78, 151
Willow carrot aphid 159
Wireworms 22, 108, 138, 158

Y

Yellow rust 15, 31, 115, 137, 140
   slime mould 30, 140

INDEX OF LATIN NAMES

A

Aclypea opaca 68, 149
Actinomycetes spp 147
Acyrthosiphon pisum 45, 143
Agriolimax agrestis 28, 36, 141
Agriotes lineatus 22
  spp 22, 108, 138
Agrobacterium tumefaciens 58, 146
Agropyron repens 137
Agrotis segetum 70, 86
  spp 110, 149
Alternaria brassicae 79, 152
  brassicicola 79
  solani 102, 103, 157
  spp 10, 49, 55, 56, 145
Amaurosoma (Cleigastra) flavipes 30
  spp 140
Amelia (Tortrix) paleana 32, 140
Anguina tritici 20
Apanteles glomeratus 87, 153
Aphis fabae 48, 67, 144, 149
Apion aestivum 142
  apricans 38, 142
  flavipes 38, 142
Ascochyta hortensis 46, 144
  imperfecta 41, 142
  pisi 44, 143
  spp 144
Athalia rosae 85, 153
Atomaria linearis 68, 149

B

Bacterium prodigiosum 100
Berberis vulgaris 16, 137
Beta virus 2 56, 57, 146
Beta virus 4 55–7, 146
Bibio hortulanus 24
  spp 139
Biston zonarius 70
Blaniulus guttulatus 109, 158
Botrytis cinerea 44, 47, 64, 120, 143, 144, 148, 161
Bourletiella hortensis 66
Brassica virus 1 74
Brevicoryne brassicae 81, 152
Bruchus pisorum 45

C

Calocoris norvegicus 66, 80
  spp 152
Cassida nebulosa 69, 149
  nobilis 69, 149
Caveriella aegopodii 159
Cephus pygmaeus 23, 138
Ceratophorum setosum 47, 144

Cercospora beticola 63, 148
  concors 102, 157
Cercosporella brassicae 78, 151
  herpotrichoides 11, 135
Ceutorrhynchus assimilis 84, 152
  pleurostigma 84, 153
  quadridens 84, 152
Chaetocnema concinna 68, 149
Chlorops pumlionis 117, 160
Cladosporium spp 10, 15, 135
Claviceps purpurea 13, 29, 136, 140
Cleigastra (Amaurosoma) flavipes 30, 140
Colaphellus sophiae 85, 153
Colletotrichum lini 111, 159
  trifolii 40, 142
Contarinia loti 41, 143
  medicaginis 43, 143
  nasturtii 75, 88, 151, 153
  spp 139
  tritici 24, 139
Corticium solani 11, 60, 104, 136, 147, 157
Corynebacterium rathayi 30, 40
  sepedonicum 119, 161
Crepidodera ferruginea 22, 138
Cystopus candidus 78, 151

D

Dactylis spp 138
Dasyneura (Oligotrophus) alopecuri 31
  brassicae 84
  onobrychidis 43
  spp 143
Depressaria spp 110, 159
Ditylenchus destructor 106, 119, 158
  dipsaci 20, 37, 42, 65, 80, 138, 141, 143, 148, 152
Dolerus spp 117, 160

E

Epichloe typhina 29, 140
Erioischia brassicae 89, 154
  floralis 89, 154
Erysiphe graminis 13, 29, 115, 118, 136, 140, 160, 161
  polygoni 35, 44, 78, 141, 143, 151
Euonymus europaeus 67
Euphorbia cyparissias 45, 143
Eurydema oleracea 80
  spp 152
Euxua spp 149

F

Festuca spp 138
Forficula auricularia 36, 66, 141

Fusarium coeruleum 103, 157
  nivale 12, 136
  roseum 103
  spp 12, 157
  sulphureum 103

G

Gaeumannomyces (Ophiobolus) graminis 10, 135

H

Haplodiplosis equestris 19, 137
Helicobasidium purpureum 64, 104, 109, 148, 158
Helminthosporium spp 118
Heterodera avenae 20, 30, 137, 140
  rostochiensis 106, 158
  schachtii 65, 148
Heterosporium phlei 118
  spp 118
Hydrellia griseola 26, 117, 139, 160
Hydroecia micacea 70, 108, 149, 158
Hylemya florilega 48, 144
Hypera nigrirostris 38, 142
  variabilis 42, 143

J

Jaapiella medicaginis 43
  spp 143

K

Kabatiella caulivora 34, 141
Kakothrips robustus 45, 143

L

Laspeyresia nigricana 46, 144
Lema melanopus 26, 30, 139
Leptinotarsa decemlineata 107, 158
Leptohylemyia coarctata 26, 139
Leptopterna dolobrata 32, 140
Leptosphaeria nodorum 160
Limothrips spp 21, 138
Luperina testacea 32, 140
Lygocoris pabulinus 108

M

Macrosiphum (Sitobion) avenae 21, 116, 160
Mamestra brassicae 70, 149
  pisi 43
  trifolii 70, 149
Mastigosporium rubricosum 118, 161
  spp 29, 140
Mayetiola destructor 32, 140
  spp 30, 31, 140

Melampsora lini 112, 159
Meligethes aeneus 83
  spp 152
Meloidogyne hapla 120, 161
Melolontha hippocastani 69
  melolontha 22, 69, 108, 149, 158
Mesapamea secalis 23, 138
Metopolophium dirhodum 116, 160
Microtus agrestis 28, 139
Myzus persicae 67, 146, 148, 149

N
Noctua spp 149

O
Oligotrophus spp 140
Oospora pustulans 105, 158
Ophiobolus (Gaeumannomyces) graminis 135
Oscinella frit 25, 139
Otiorrhynchus ligustici 43, 143

P
Passer domesticus 28, 139
Pectobacterium carotovorum 76, 100, 151, 156
Pectobacterium carotovorum var. atrosepticum 99, 156
Pegomyia betae 70, 149
Peronospora farinosa 61, 147
  parasitica 78, 151
  trifoliorum 41, 142
Philopedon plagiatus 84
Phoma betae 56, 60, 62, 147
  extigua 119
  lingam 79, 152
  solanicola 119, 161
Phyllopertha horticola 69, 149
Phyllotreta nemorum 82, 152
Phytonomus arator 22
Phytophthora infestans 102, 103, 157
Pieris brassicae 87, 153
  rapae 87, 153
Plasmodiophora brassicae 77, 151
Plutella xylostella 86, 153
Poa spp 138, 140
Polyspora lini 111, 159
Polytrinchium trifolii 34, 141

Pseudocercosporella herpotrichoides 11, 135
Pseudomonas coronafaciens 9, 13, 135
Pseudopeziza medicaginis 41, 142
Psila rosae 110, 159
Psylliodes chrysocephala 83, 152
Puccinia coronata 15, 16, 31, 118, 137, 140, 161
  graminis 15, 137
  hordei 15, 137
  poarum 31, 140
  recondita 15, 31, 140
  striiformis 15, 115, 137, 140, 160
Pyrenophora avenae 14
  graminea 114, 160
  gramineum 14
  spp 14, 136, 161
  teres 14
Pythium debaryanum 60, 147
  spp 62, 147
  ultimum 147

R
Ramalaria betae 63, 148
Rhamnus cathartica 16, 137
Rhizoctonia carotae 120, 161
  cerealis 11, 136
Rhopalosiphum padi 116, 160
Rhynchosporium secalis 4, 14, 136
  spp 161

S
Sclerotinia sclerotiorum 109, 158
  spp 144
  trifoliorum 35, 41, 141, 142
Septoria avenae 9, 135
  linicola 112, 159
  nodorum 115
  tritici 13, 136
Siteroptes graminum 32, 140
Sitobion (Macrosiphum) avenae 21, 116, 160
Sitodiplosis mosellana 24
Sitona linealus 39, 46
  spp 142, 144
Solanum virus 14 96
Spondylocladium atrovirens 105, 158

Spongospora subterranea 101, 105, 156, 158
Stemphylium radicinum 109, 158
  sarciniforme 34, 141
Steneotarsonemus spirifex 27, 139
Streptomyces scabies 59, 105, 147, 157
  spp 59, 119, 147
Subcoccinella 24-punctata 68, 149
Synchytrium endobioticum 101, 156

T
Thanatephorus cucumeris 104
Thielaviopis 144
Thrips angusticeps 21, 66, 81, 138, 148, 152
Tilletia caries 17, 137
Tipula paludosa 24, 89
  spp 139, 153
Tortrix (Amelia) paleana 32, 140
Trioza apicalis 110
Typhula brassicae 64, 79, 148
  trifolii 41

U
Urocystis occulta 17, 137
Uromyces betae 63, 147
  pisi 45, 143
  trifolii 36, 141
Ustilago avenae 18
  bromivora 30
  hordei 18
  nuda 18, 137
  perennans 30
  spp 140
  tritici 17, 137

V
Verticillium albo-atrium 40, 142
Virus A 97, 155
  G 95
  S 97
  X 97, 156
  Y 95, 97, 98, 156

X
Xanthomonas campestris 75, 76, 151

ACKNOWLEDGEMENTS

Thanks are due to Mr J. Lamont for his help in checking over the text and to Miss A. Hurn for typing the manuscript.

The artwork for the colour plates was prepared by Ingeborg Frederiksen and Ellen Olsen. Photographs were supplied by: BASF Ltd (Plate 115 F & G); J. Begtrup and B. Engsbro (Plate 113 C); E. Hellmers (Plate 118 E & G); A. Jensen (Plate 120 F); A. From Nielsen (Plate 119 C); V. Smedegård Petersen (Plate 113 B).